A Quaker Approach to Research:
*Collaborative Practice and
Communal Discernment*

Quaker Institute for the Future Pamphlet Series

1— *Fueling our Future: A Dialogue about Technology, Ethics, Public Policy, and Remedial Action,* coordinated by Ed Dreby and Keith Helmuth, edited by Judy Lumb, 2009.

2— *How on Earth Do We Live Now? Natural Capital, Deep Ecology, and the Commons,* by David Ciscel, Barbara Day, Keith Helmuth, Sandra Lewis, and Judy Lumb, 2011.

3— *Genetically Modified Crops: Promises, Perils, and the Need for Public Policy,* by Anne Mitchell, with Pinayur Rajagopal, Keith Helmuth, and Susan Holtz, 2011.

4—*How Does Societal Transformation Happen? Values Development, Collective Wisdom, and Decision Making for the Common Good,* by Leonard Joy, 2011.

5—*It's the Economy, Friends: Understanding the Growth Dilemma,* edited by Ed Dreby, Keith Helmuth, and Margaret Mansfield, 2012.

6—*Beyond the Growth Dilemma: Toward an Ecologically Integrated Economy* edited by Ed Dreby and Judy Lumb, 2012.

7—*A Quaker Approach to the Conduct of Research: Collaborative Practice and Communal Discernment,* by Gray Cox with Charles Blanchard, Geoff Garver, Keith Helmuth, Leonard Joy, Judy Lumb, and Sara Wolcott, 2014.

8—*Climate Change and Food Security: A Quaker Approach to the Prevention of Violence,* by Judy Lumb, Phil Emmi, Mary Gilbert, Laura Holliday, Leonard Joy, and Shelley Tanenbaum, 2014.

A Quaker Approach to Research:
Collaborative Practice and Communal Discernment

Gray Cox
with Charles Blanchard, Geoff Garver
Keith Helmuth, Leonard Joy, Judy Lumb, and Sara Wolcott

Quaker Institute for the Future Pamphlet 7
Quaker Institute for the Future 2014

Copyright © 2014 Quaker Institute for the Future

Published for Quaker Institute for the Future by *Producciones de la Hamaca*, Caye Caulker, Belize <producciones-hamaca.com>

ISBN: 978-976-8142-57-3

A Quaker Approach to the Conduct of Research: Collaborative Practice and Communal Discernment is the seventh in the series of Quaker Institute for the Future Pamphlets:

Series ISBN: 978-976-8142-21-4

Quaker Institute for the Future Pamphlets aim to provide critical information and understanding born of careful discernment on social, economic, and ecological realities, inspired by the testimonies and values of the Religious Society of Friends (Quakers). We live in a time when social and ecological issues are converging toward catastrophic breakdown. Human adaptation to social, economic and planetary realities must be re-thought and re-designed. *Quaker Institute for the Future Pamphlets* are dedicated to this calling based on a spiritual and ethical commitment to "right relationship" with Earth's whole commonwealth of life.

Quaker Institute for the Future
<quakerinstitute.org>

Producciones de la Hamaca is dedicated to:
—Celebration and documentation of Earth and all her inhabitants,
—Restoration and conservation of Earth's natural resources,
—Creative expression of the sacredness of Earth and Spirit.

Contents

Preface vii
Introduction ix

CHAPTER ONE
Quaker Traditions of Communal Discernment
Vision and Theory of Quaker Traditions 2
Quaker Practice of Communal Discernment 5
Quaker Meeting for Worship for the Conduct of Business 7
Threshing Sessions 8
Experience of Threshing Meetings 8
Clearness Committees 9
Quaker Dialogue 9
Setting Priorities 9
Models of Quaker Action 10

CHAPTER TWO
Meeting for Worship for the Conduct of Research
A Quaker Approach to Research 13
Clearness Committees for Research 14
Experience with a Clearness Committee for Ethnographic
   Research 15
Research Seminars and Similar Workshops 17
Summer Research Seminar Experience 20
Circles of Discernment 21
Reflections on *Fueling Our Future* 23
Reflections on *Beyond the Growth Dilemma* 23
Book Project: The Experience of Writing *Right Relationship* 24
Lessons from a Quaker Approach to Research 28
Summary 29

CHAPTER THREE
Comparing Traditions of Communal Discernment
Collective Discernment in Indigenous Cultures 32
Spiritually Framed Traditions of Communal Discernment 36
Extending Quaker Practice to Secular Settings 37
Experience of Quaker Process in a Secular Setting 39
Quaker Process in a Biology Graduate Program 43

CHAPTER FOUR
Philosophical Issues Concerning Communal Discernment
Communal Discernment as a Method of Discovery 47
Communal Discernment as a Method of Justification 47
Teleology in Quaker Communal Discernment 52
Design Space 55
Communal Discernment in Social Research 57
The Inward Light, Intellect, and Imagination 62
Summary 70

CHAPTER FIVE
Closing 72

Endnotes 74
Bibliography 77

Preface

A Quaker Approach to Research: Collaborative Practice and Communal Discernment grows out of a decade of experiments by Quaker Institute for the Future (QIF) employing Quaker processes of communal discernment in research in the context of public policy, academic study, and community-based research. This pamphlet is itself the product of a collaborative process which began in 2003 with discussions of practicing "Meeting for Worship for the Conduct of Research" to use a collaborative, communal approach instead of the current competitive, argumentative approach to research. During the last decade, people associated with QIF seminars and projects have engaged in reflective discussions, worshipful meditations, and practical experiments exploring this question. The side bars in this pamphlet offer illuminating testimonies from a number of their experiences.

In the QIF Summer Research Seminar of 2013, I shared a paper proposing a Circle of Discernment on Quaker Epistemology and invited Friends to join me. The task was taken up by a group initially consisting of Judy Lumb, as convener, Charles Blanchard, Geoffrey Garver and myself. We were later joined by Leonard Joy and Sara Wolcott. For me, our work on this pamphlet has been an especially rich, rewarding, collaborative process in which all participants contributed in major ways, not only to the discernment process but also to the substantive text as well. Sections for which contributing authors were primarily responsible are marked in the text but the entire pamphlet was also reviewed and reformed by their ideas, experiences and suggestions. The text attempts to accurately reflect the practice and findings of those who have worked with QIF.

Members of the Religious Society of Friends (Quakers) are often called to help generate knowledge, wisdom, and systematic insight that can inform public policy and our practical efforts to transform society. I hope this pamphlet will inform our research efforts in ways that will enable us all to live more fully in that life and power that leads us to treat all humans and all of Earth as a manifestation of the Divine.

This QIF Pamphlet is designed to be of interest to at least three audiences: 1) Quakers who are interested in ways their experiences of worship and faith could inform their research or other work in the world, 2) Physical and social scientists from other faith traditions who are interested in how their spiritual lives might be more fully integrated with their work and 3) Researchers who are interested in how Quaker practices of communal discernment, if viewed in secular terms, might be relevant to enhancing the forms of collaboration they practice in their work.

Readers are likely to encounter language that describes experiences or expresses ideas in terms that are unfamiliar or are in tension with their accustomed ways of thinking. Some readers may find sections of Spirit-centered language either puzzling or problematic. Other readers may find language that departs markedly from their own faith tradition in ways that may not do justice to their experiences of the Divine.

Because the intent here is to encourage dialogue between diverse audiences, my fellow authors and I hope that readers will enter into dialogue on the themes developed in this pamphlet and find their openness well rewarded. We hope readers will find the idiomatic language of the Religious Society of Friends translated into terms that may advance insights for a wide range of mainstream researchers and policy analysts.

I am especially grateful to the contributing authors and editors who did so much to give this pamphlet substance and style: Charles Blanchard, Geoff Garver, Keith Helmuth, Leonard Joy, Judy Lumb, and Sara Wolcott.

I am deeply grateful to the others who shared material offered in sidebars reflecting on their experiences in collaborative research including, Pamela Haines, Laura Holliday, Robert McGahey and Stanford Searl.

Very helpful and very much appreciated readings and critical reviews of the pamphlet were provided by Barbara Day, Stanford Searl, Shelley Tanenbaum, and Vince Zelazny.

I am especially grateful to, and admiring of, Judy Lumb for her stellar work in coordinating our meetings and work, and overseeing production and publication of this pamphlet.

—Gray Cox
Bar Harbor, Maine
May 23, 2014

Introduction

Since its founding in 2003, the Quaker Institute for the Future (QIF) has been experimenting with ways of doing Spirit-led research using variations on Quaker practices of communal discernment first used in the mid-1600s. Quakers hold that there is that of God in everyone, and that revelation is a continuing process. For Quakers, worship is the response to an awareness of God within. The practice of worshiping attentively and openly in the silence can enable a community to discern Truth that comes in messages and leadings that can and should guide our lives. Because Quakers sought from the beginning to be governed, not by individual men or women, but by leadings of the Divine from the Spirit-led wisdom of the community as a whole, they developed a practice of "Meeting for Worship for the Conduct of Business."

The practice has come to include a rich variety of specific methods for communal discernment. Members of QIF have sought to extend these practices explicitly into the realm of public policy and academic studies by experimenting with methods for "Meeting for Worship for the Conduct of Research." These experiments have included the annual Summer Research Seminars, a major collaborative book, *Right Relationship: Building a Whole Earth Economy*, and a series of pamphlets, some of which were written by teams forming "Circles of Discernment" on topics such as *Fueling Our Future, Genetically Modified Crops*, and *Beyond the Growth Dilemma: Toward an Ecologically Integrated Economy*.[1]

The aim of this pamphlet is to describe 1) the vision, theory, and traditions of practice inspiring a Quaker approach to research; 2) experiments with specific methods used; 3) initial results and findings; and 4) the key challenges and puzzles that remain. It further aims to explore the relevance of Quaker process when the participants are not Quaker or even religious.

Chapter I begins with a description of the vision, theory, and tradition of practices that emerged among early Quakers in the 1600s and have been refined and extended in a variety of decision-making contexts over the last 350 years. Chapter II provides a more detailed account of these procedures and practices as applied to research with which QIF has experimented in its first ten years.

Chapter III compares other traditions with these Quaker practices of communal discernment in research, policy analysis and collective decision making. There is much to be learned through dialogue with other faith-based traditions. Further, it is useful to consider how Quaker practices may be modified and adapted in secular settings.

Chapter IV explores some of the philosophical issues and challenges that are raised by the very idea of having a "Quaker Epistemology" or way of knowing and the distinctive assumptions made about the process of research, the norms and criteria for knowledge, and the nature of reality. Especially challenging are the historic splits between church and state, and between faith and reason. This chapter does not provide final resolution to the important methodological and metaphysical issues raised but tries to frame them in reasonable and useful ways to facilitate ongoing dialogue amongst Quakers and others.

Chapter V closes with some reflections on the future of collaborative and communal research.

CHAPTER I
Quaker Traditions of Communal Discernment

This chapter aims to introduce the Quaker tradition of practices associated with Meetings for Worship for the Conduct of Business and other models of Quaker action. This will provide a background for the discussion in Chapter II of QIF experiments with variations on these practices in Meetings for Worship for the Conduct of Research.

The manner of worship of "unprogrammed" Friends[2] is silent waiting upon the Spirit. Friends may speak out of the silence when moved by the Spirit. Friends are organized according to how often that group gathers to conduct business. For example, "monthly meetings" are local groups that gather for worship at least once a week and for the conduct of business each month. "Quarterly meetings" and "yearly meetings" are composed of monthly meetings and meet quarterly and annually. The business of the meeting is conducted in the same manner as worship with speaking out of the silence, and is called a "Meeting for Worship for the Conduct of Business."

The Religious Society Friends, at all levels of organization, is governed through communal discernment by the body of participants without any hierarchy authorized to provide a definitive, official formulation of Quaker practice. However, a variety of articles, pamphlets and books have been written studying aspects of the process and most yearly meetings have published books of "Faith and Practice" generated by group discernment of representatives from the entire yearly meeting to document their practice and provide guidance for monthly meetings.[3]

Vision and Theory of Quaker Traditions

Quakers emerged in the mid-1600's in Britain, in the midst of a civil war and social upheaval that has been characterized as "a world turned upside down."[4] Amongst the many that came forward urging religious reform and social change, were groups like the Levelers, the Diggers, the Ranters, and the Fifth Monarchy Men. One group, who called themselves the "Seekers," did not claim to have found the Truth but believed that if they gathered patiently in silent worship, waiting on the Divine, God would provide insight and guidance. This activity of silent, expectant, worshipful waiting and listening was incorporated by the early Quakers into their practice. But the early Quakers did not feel they waited in vain. In fact, they had the strong experience of encountering a presence of the Divine. They described this experience as the "Inward Light," "the Seed," "that of God" in everyone, and "Christ who has come to teach his people himself."

In choosing to call themselves the "Religious Society of Friends of the Truth," early Quakers were witnessing to the presence of a Power that can provide openings to a more explicit, clear, self-consistent, accurate, full understanding of reality, and has a transformative power over our lives and those of others. They found methods for turning toward this power and tuning in to it. These were methods that could be learned, that were open to all, available at any time, practices that could be used to structure all of life. The practice of these methods and the experience of this presence of the Divine provided the central, defining feature of their faith and practice. It gave rise to a very distinctive, non-creedal, understanding of the nature and meaning of faith and revelation, and a distinctive, holistic, and process-centered view of reality.

The Quaker community is one without a creed. Faith is understood, not as a set of beliefs that are adopted, but as a series of experiences that provide queries and leadings that guide. The Quaker faith is not a commitment to a list of fixed statements, such as the Nicene Creed, but consists instead in the sharing of experiences of the Divine that are encountered in worship through the presence of "that of God in everyone." The commitments and concerns of Quakers are best understood in terms of historic testimonies and queries that "speak to" them as specific individuals and communities. For example, traditional Quaker concerns about peace are not laid out in a code of abstract, impersonal principles that provide orders everyone ought to obey. Instead, the tradition offers personal witnesses of people like John Woolman, who found himself led to travel for a time among his Native American brothers and sisters. The tradition offers questions that give pause and asks each person to consider with care: "Do you live in the virtue of that life and power that takes away the occasion of all wars?"[5]

For Quakers, revelation is a continuing process, ongoing in the world today and available—through that of God in everyone. This means that everyone is related to the world, to each other, and to the Divine in ways that cannot be described well in atomistic and static terms but need, instead, to be understood in holistic and emergent ways. This applies to the ways in which rational ideas emerge from felt impulses and leadings. It applies as well to ways in which individuals and communities find the meanings they share, the truths they encounter, and the identities that define them in emergent, interconnected, holistic ways. Such holistic emergence occurs in the context of the interconnected aspects of everyone and everything involved in patterns of collective intelligence. If the understanding of social reality is emergent and holistic, then no sharp and radical distinction can or should be made between the realms of the spiritual and the secular, between faith and reason, or between religion and science.[6]

Quaker attitudes toward concerns—and the process by which they are addressed—are rooted in fundamental beliefs about truth, meaning, reason, and the self.

Quakers view *truth* as something that happens; it occurs. Truth is not a dead fact that is known; it is a living occurrence in the depth of worship or in bearing witness and "speaking truth to power." But such events or occurrences of truth are not merely ephemeral, subjective phenomena. They are grounded in and continuations of the witness of earlier Friends and others who have found unity in a reality that is external, coming from an independent and objective source. The origins of this view lie in the conception of the Christ as a living truth that comes as a Presence whenever two or more are gathered.

Quakers experience this Presence as something available to all people and as a source of continuing revelation. Non-Christian

Friends experience this kind of living Truth and express it in other terms that do not refer to Jesus of Nazareth—or hold close to Christian vocabulary. Some modern Quakers express their experiences of God in the language of energy and resonate with concepts from modern physics that have developed around revelations that atomic structures are mostly empty of matter but filled with immense energy. Other Quakers may simply feel that the quintessential Quaker insight—"that of God in everyone" or the "Inward Light"—can be seen as an expression of an even more ancient and enduring intuition.

Michael Sheeran notes that *"Unlike other faiths, Quakerism builds all on the experience of the gathered meeting. Together Friends experience something beyond themselves, superior to the human pettiness that marks ordinary life. One may find in this experience the Spirit of Christ, another the Divine Person, a third the force behind the universe. No matter how they explain the experience to themselves, the event in which they share is paramount. They stand in awe before it, finding that it dominates their conduct as they meet together to make a decision. And the event demands that, in reaching that decision, they should sacrifice self-interest and seek after a higher truth than what they have individually achieved."*[7]

Meaning for Quakers is a communal process. It is like a dance; there are individual bodies moving, but one dance is occurring. When Quakers gather the sense of a meeting to reach a decision, they do not speak of counting heads to see who means what. Instead, they ask what they, collectively, mean. Quakers may say many different things, and yet somehow speak with one voice. George Fox, one of the seventeenth century founders of the sect, said that this is the voice of Christ who "has come to teach His people Himself."[8]

Feeling and **reason** are viewed as interactive with one another. The Quaker view of this is analogous to one implicit in the language of many professionals, executives, and activists who make use of hunches, intuitions, gut feelings, and other experiences of emotional responses to situations—but who work to carefully discern just why they feel a certain way about some option they are considering.

In reflecting on the Quaker experience of this in the context of silent worship, Stanford Searl has noted the way "the kinesthetic pulsations of one's body remembers, becomes aware of, carries some of the experiential meanings that may be beneath the surface of ordinary language or consciousness, but somehow bubble-up at times, possessing one with feelings, sensations and experiences that may or may not have words attached."[9]

In the course of becoming explicit about such bodily and emotional responses to the world, they discover that these responses typically express judgments that have reasons. And discerning the nature and merit of such reasons implicit in their feelings is a central part of the reasoning process by which they make cultivated, rational, well grounded, wise decisions. Similarly, Quakers speak of feeling "stirrings of the Spirit," "leadings," "finding themselves moved or led," and "having tenderings of the heart." These responses implicitly express insights and reasons that provide the basis for rational discernment of the truth about the concerns they are addressing. Similarly, some writers have argued that there are characteristic "women's ways of knowing" that likewise rely on a continuity of feeling and reason, instead of an insistent dry and unemotional rationalization.[10]

The *self* is viewed as inherently social and transitional, as becoming. However, Quakers find that the experience of communal discernment presents a reality that demands that both the communal character of the self and the radical worth of individuals be acknowledged. Free and responsible individuals are addressed and conjoined in the paradoxical reality of relations between "I" and "thou" (or "me" and "thee"). It would be a mistake to think that Quaker tradition pictures the self as merely social, as though it were a secular artifact created solely by social forces. At the heart of the community in which Quakers participate is a spirit, a spirit that grows out of each one and yet also grows into each one. Historically, the paradoxical character of this notion was captured by the Christian metaphor that views individuals as distinct members of a unified body of Christ. This view of the self means that thoughts and actions are guided, not by what is best for individuals considered separately from the group, but rather as interdependent parts of a whole.

Quaker Practice of Communal Discernment

Quaker communal discernment is a practice of participatory decision making that does not employ voting. Instead, concerns are raised, discussed, and subjected to prayerful reflection until unity is reached. The process extends beyond consensus and is sometimes termed "sense of the meeting."

This is an open process that leads to new insights and perspectives that allow a "way to open" for the community to discern a truth. It is an activity born of shared commitments and concerns rooted in a coherent set of ideas as basic as the nature of meaning and truth, and practiced in a disciplined way.

The actual process of collective decision-making can be understood in terms of five phases, though it is important to emphasize that these overlap in practice, and it is common to move back and forth among them. The phases are as follows:

Quieting Impulses can be understood as the stage of "centering down" or "entering the Silence" in which one seeks to let go of the push and pull of desires and fears of everyday life.

Addressing Concerns occurs when, once having entered the silence, one can seek to listen attentively for messages that reflect the "small still voice." Leadings or callings seem to come, not from personal desire or fear, but from some higher source. These provide guidance or a concern under which one should labor to try to discern the Truth and see how to live in right relationship with others.

Exploring Responses and Gathering Shared Insights is the phase of the process that involves seeing things from different perspectives, trying to collaborate in understanding them in a more holistic, clear, explicit, accurate, coherent and complete way.

Finding Clearness is the phase at which the exploration aims and to which it is drawn, one where sharing has allowed differences to find an inclusive, compelling unity.

Bearing Witness is the phase in which such unity of understanding leads to action that embodies it in the world. It is encapsulated in the Quaker story about the visitor who came to a Meeting for the first time and half way through the hour of silence turned to his neighbor and asked, "When does the Service begin?" To which the reply was: "The service begins when the worship is over."

These five can occur in worship, in meeting for worship for business, when hassling things out with a spouse or employer, or while working as part of a social movement. They can occur through all of life. Focusing on one of the five at a time makes them seem like steps. Alternatively, at any given moment all five may be present as integral parts of a decision-making process.[11]

The Meeting for Worship in which people gather for an hour or more in silence to wait expectantly, listening to that of the Divine, is the central practice of Quakers. But they have adapted it to a variety of meeting formats that serve distinctive functions. These include the Meetings for Worship for the Conduct of Business, Threshing Sessions, Clearness Committees, Quaker Dialogues and a variety of formats used in action and social change contexts, such as setting priorities for the Friends Committee on National Legislation, Listening Projects, and Imaging a World Without Weapons workshops. The remainder of this chapter describes these variations.

Quaker Meeting for Worship for the Conduct of Business

A Quaker Meeting for Worship for the Conduct of Business begins with a period of centering silence. Some meetings have a tradition of a query or quotation being read during that silence for the reflection of the community. The Presiding Clerk then breaks the silence and presents an agenda. Upon approval of the agenda, the meeting proceeds through the business as presented. Decisions are made by unity or the "sense of the meeting." After some discussion, the Clerk may sense there is unity, verbalize the potential decision, and ask if anyone has unresolved concerns. Those not in agreement will state their concerns. If concerns have been addressed and a written minute has already been prepared, or is not needed, Friends in agreement say "approved" when the Clerk requests approval. If there is a need to prepare or revise a written minute, the gathering goes into silence while the Recording Clerk drafts the minute for approval of those gathered.

A common practice is to call for silence when emotions are running high or when people find it easier to speak at length than to listen in depth. Often, out of that silence, a convergent solution emerges that is quite different from those previously proposed.

Still, some individuals may find themselves unable to join in unity with a decision, but they are in agreement that it is better for the group to go forward. They might choose to "stand aside," sometimes asking to be "minuted" as standing aside.

Alternatively, if there are important differences on an issue, it may seem wise to "season" it by putting it aside until a later Meeting for Worship for the Conduct of Business. A few Friends might gather to "labor over it." What can make the labor difficult, but most rewarding, is to select precisely those Friends who are most in opposition to each other to labor together with the aim of giving birth to a vision with common ground. These and other aspects of Quaker process and tradition provide potential experiments in collaborative efforts at research.

Threshing Sessions

A Threshing Session is called for the entire Meeting on occasions when an issue has emotional undertones, broad implications for the Meeting, and is difficult to resolve. The objective is not to make any decisions, but to air feelings, as threshing separates the wheat from the chaff. It is usually helpful to divide the larger group into smaller groups for a period of time and then to come back to the larger group. Once feelings are expressed and heard, a deeper level of understanding can be attained. Often it is then possible to make appropriate decisions at a subsequent Meeting for Worship for the Conduct of Business.

Experience of Threshing Meetings

—Judy Lumb

I was a sojourning member of 15th Street Meeting in Manhattan, New York City in 1980-81. At the November 1980 Meeting for Worship for Business an announcement was made in the facilities report that the meetinghouse would be used for a lesbian wedding under the care of another monthly meeting. No action was required as this was merely an announcement, but the gut reactions expressed surprised everyone, even those speaking. The Clerk scheduled a threshing session three months later. Every First Day, he explained that a threshing is designed to separate the wheat from the chaff, to air feelings, but not to make any decision. The threshing session began with small group worship-sharing sessions focused around each Friend's first awareness of homosexuality and how their feelings had changed over the years. When the entire group reconvened in the final worship, there were amazing messages about healing that had happened. Elderly people spoke about how their current feelings arose from teenage experiences that they had never told anyone. It was a very effective consciousness-raising event in regard to homophobia.

Two threshing meetings were held around Atlanta Friends Meeting's decision to sell Quaker House, which was a mansion purchased in the late 1950s with funds from Philadelphia Yearly Meeting for witness in the Civil Rights movement. It was one of the few places that an integrated gathering could take place, so there was a lot of history associated with the building. But by the 1980s Atlanta Monthly Meeting had out-grown the space. Because it was in an historical district, there was no possibility of renovating to make a more appropriate meetinghouse. To facilitate the decision, two threshing sessions were held in May of 1988 to process everyone's feelings about Quaker House. Then it was possible to make the decision to sell Quaker House at the Meeting for Worship for Business in June 1988.[12]

Clearness Committees

Clearness Committees use a variation on the Meeting for Worship for the Conduct of Business to help an individual or group to test a plan, proposal, or leading they have to travel, marry, or undertake some other action. A Clearness Committee may also be used to settle conflicts between individuals. A clerk convenes a small group of three to six people. They settle into silence out of which the individual asking for clearness explains the leading or situation. The others listen and speak out of the silence their reflections on the issue, helping the individual or individuals come to clearness.[13]

Quaker Dialogue

Quakers have developed a number of specific ways for carrying out the practice of communal discernment that help the process along. One is to open up discussion on an issue with a "Quaker Dialogue" in which a theme or topic is addressed by each person in the circle, one at a time, starting with some silence to allow everyone time to center down and speak in a more Spirit-led way and, very importantly, allow a good measure of silence between sharings to encourage responses to divine leadings rather than shallow give-and-take chatter about each other's words.

Setting Priorities

Friends have developed practices to conduct business that extend to the life of the Religious Society of Friends at large and include methods of inter-visitation between meetings, using working groups to gather concerns and insights from Monthly, Quarterly or Yearly Meetings, or holding consultations amongst Friends at large.

One example is the process developed by the Friends Committee on National Legislation (FCNL) for discerning legislative priorities. It involves a year-long activity in preparation for each newly elected Congress. All communities of Quakers throughout the United States, including the very liberal, very conservative, and evangelical Friends are asked to consider their legislative priorities, how they want FCNL to focus their lobbying efforts on behalf of Friends. Minutes from these grassroots, worship-based discussions are then gathered and reviewed by the FCNL Policy Committee in a series of worship-based discernments. The Policy Committee prepares a draft document that reflects the priorities identified by the participating Quaker Meetings, groups, and individuals. This document is then reviewed and refined in a similar way at an annual meeting of the FCNL General Committee where 200 or more delegates from Quaker Meetings and organizations from around the country provide a final testing and approval of the priority-setting document. The procedures of this discernment have become highly nuanced because the aim is not to simply nail down a document that the General Committee can approve, but to frame a message that includes the collective, Spirit-led discernment of those who have taken part in the process but are not present—and of that independent, objective Truth to which they are all seeking to attend. This document is then used to guide the lobbying efforts by the staff and volunteers of FCNL for the next Congress.

Models of Quaker Action

Quakers have developed methods that provide models of action that draw on and extend the methods of discernment at the core of the tradition. They include, for example:

- The Listening Project, developed by Herb Walters and other Quakers, uses "trained volunteers to conduct one-on-one interviews that address local and sometimes national or international issues. Interviewers take time to build trust and understanding, so that people interviewed can go deeper into their fears, hurts, hopes, needs, feelings and ideas. ... As citizens begin to understand that their feelings, opinions and actions can matter, they respond in dynamic ways. Some offer creative ideas and solutions. Some take the next step to action or leadership. Thus the Listening Project can be an important step toward individual and community empowerment."[14]

- Elise Boulding and Warren Ziegler developed a workshop for "Imaging a World Without Weapons." This involves a seven-step, multi-day process in which a wide variety of imaging methods aid participants in discerning the kind of future they feel most called to work toward. It draws on methods for creating "deep listening" and provides ways for individuals or groups to discern leadings and articulate visions.[15]
- The Alternatives to Violence Project has developed methods to "use the shared experience of participants, interactive exercises, games and role-plays to examine the ways in which we respond to situations where injustice, prejudice, frustration and anger can lead to aggressive behaviour and violence."[16]
- Rex Ambler has explored and interpreted the inner spiritual practices of early Friends and developed instructions to apply them in a variety of modern contexts.[17]
- The work of John Woolman provides another interesting Quaker model. His journals of his travels, his work on the "Plea for the Poor," and his reflections on slavery might be considered models of Quaker research in economics. In a 2008 reflection on the travels of John Woolman, Friends Ruah Swennerfeldt and Louis Cox walked the Pacific Coast of the U.S., sharing their concerns and challenging Friends on their "Peace for Earth Walk."

The methods described here require practice in order to use them wisely and well. They have their counterparts in a variety of other spiritual traditions. They are not mere gimmicks or tricks to be imitated blindly, nor are they the proprietary knowledge of a single religious community. Instead, they are practices that are meant to be pursued with intent to be Spirit led, seeking Truth in openness to all the possibilities the universe may offer, including the insights and teachings of other traditions. The next chapter discusses these methods in the context of research.

Chapter II
Meeting for Worship for the Conduct of Research

For Quakers who seek to ground their actions in their spiritual lives, what is the relationship between spiritual discernment and the practice of disciplines that generate knowledge used to define and guide policy for Quaker communities and individuals? How might Quakers and researchers from other faith traditions, as well as from secular, agnostic, and atheist perspectives, find common ground and shared insight in the assumptions that frame their methods of research and the nuances of practices in which they pursue theoretical knowledge and practical wisdom? Might it be that Friends are called to explore the development of research methods that are Spirit-led and provide a way of living out our historic role as "Friends of the Truth,"[18] who are guided by that Power and Presence that has come to "teach his people himself"?[19]

The application of Quaker traditions to research seems fairly straightforward. Research is a kind of human activity. Why not treat it like any other and approach it using Quaker collaborative methods? In many ways that has been the approach adopted in QIF and it has met with significant success. This chapter will sketch some of QIF's experiments including Clearness Committees, Research Seminars and Workshops, Circles of Discernment, and a book project.

In the world today research is a basic foundation of the social order. Studies in economics, anthropology, biology, psychology, philosophy, climate science, and other fields are used to understand the nature and source of social and environmental problems and propose responses that are effective in serving the common good. Such studies purport to determine how the world works and how one should live in it. All too often, contemporary social, natural science, and public policy research is rooted in traditions of "publish or perish" institutional demands, territorial "turf wars," and argumentative styles of writing and debate that often make use of war metaphors to frame the "attacks" and "defenses" that structure public reasoning.[20]

Much of such research is defined by a commitment to secularism, a separation of church and state that is often extended to a separation of church and academy, and a split between faith and reason framed by a conviction that objectivity in science precludes any reliance on religious revelation or spiritual leadings. In contrast to such competitive research and the secular, often positivist, assumptions used to frame it, the Quaker process of communal discernment provides a fundamentally different approach to doing research and sharing its results.

A Quaker Approach to Research

Epistemology refers to the study of knowledge, deriving from the Greek words "episteme" (knowledge) and "ology" (the study of). Epistemology studies the criteria and methods that yield truth in science and elsewhere and that are independent of particular creeds and cultures. To many philosophers, the nature of knowledge would appear to be universal in character, so the notion of a "Quaker way of knowing" or a "Quaker epistemology" might seem obscure or even bizarre, as though there would be a Presbyterian physics or Episcopalian endocrinology.

On the other hand, to many anthropologists and other students of culture the idea might seem quite straightforward. They expect every culture and sub-culture to have its own criteria and methods for constructing their social reality and determining what is done and not done, said and not said, and true and not true in their culture. Just as there are different "ways of knowing" or epistemologies for Shinto priests, Yaqui Indians, and Mayan H'men, it would seem quite obvious to expect that Quakers would have distinctive epistemic practices that include criteria and methods for how they know.

But then the question might be: Do the Quaker methods of knowing have any normative force for others? Are the Quakers on to something? Is there anything in their particular practices for deciding what is to be done and not done, said and not said, or affirmed as true or false, that could be adopted by others doing research on public policy questions and social change? What would be the real value of these methods? To what extent might they actually help to know more or know better?

The Quaker belief that the Light of Truth can be found in each and every person may seem inconsistent with the idea that there is a special method of knowing the truth that Quakers have developed and that provides unique insight into the dimensions of truth. Because communal discernment is a human potential, it should be possible to include this process in any of the practices, technical methods, and approaches of other epistemic traditions in research. This has been the experience in the QIF Summer Research Seminars where participants bring a wide variety of disciplinary approaches. The silence is big—big enough even for polemical writings, for instance, as long as the frame in which they are listened to, and dialogued with, is the larger frame of Love rather than competition.

Clearness Committees for Research

Friends may initiate research by preparing a brief statement of research plans, and requesting a Clearness Committee from their Monthly Meeting. People taking part in QIF work in Summer Research Seminars or as members of the Board have found this an especially helpful way to begin. The standard process that Monthly Meetings use for Clearness Committees can function perfectly well in this context. A typical procedure would involve two to four members of the Meeting who are asked to meet with the person (or persons) seeking clearness.

They may be invited by that person or by the Monthly Meeting or typically some combination of the two. An initial meeting might begin with a period of silence, followed by the person seeking clarity introducing her concerns and their context. Continuing to share out of the silence, the others may offer comments, often in the form of brief queries that invite deeper reflections and openings. The seeker may express and affirm whatever insight, clarity or revelations she may wish to be tested in the response of the other members of the committee. Usually there is some closing silence to allow the process and its results to resonate. A key element in the process is deep listening, which allows tacit forms of knowledge—and doubt—to find expression. The silence, the presence of attentive others, turning to queries rather than pushing arguments, and openness to the use of metaphor, narrative and other forms of indirect or poetic language can all help to deepen listening, levels of thought, feeling, context and Spirit on which the clearness process draws.

In research contexts, it is helpful to include a mix of people on the committee who are familiar in some way with the area of research and others who are non-academic and non-research-oriented Friends. One might frame a research project as an open-ended query of the sort that Quakers have traditionally found a powerful way to invite openings, instead of a question that arises out of a theoretical framework or a political conflict. Such queries include, for example, the oft-quoted one based on a phrase of George Fox, "Do you live in virtue of that life and power that takes away the occasion of all war?" Such queries do not ask for immediate and simple answers like "yes," "no," or "maybe," but instead invite extended reflection and questioning on the meaning of the question itself.

In projects studying issues that have been controversial amongst Friends (like nuclear energy or the use of neo-classical economic theory in policy analysis of the commons), research teams have found it especially helpful to frame their work with open-ended queries that invite deep listening. Often the result of an initial clearness committee meeting may be simply clarification that will further the research.

While Clearness Committees can, in principle, just meet once with a seeker and then be "laid down," it is often very helpful for them to continue in a support role. They may meet formally at intervals to review the work undertaken and report back to the Monthly Meeting or other Quaker group from which they were formed. Or individual members may simply check in with the seeker and provide support in less formal ways.

Experience with a Clearness Committee for Ethnographic Research
—Stanford Searl, Jr.

I would like to open up an illustration about the clearness process, based upon the experience of being guided by the Peconic Bay Friends Meeting on the end of Eastern Long Island. We met monthly for two years as an official oversight and clearness committee for myself and my research project about Quaker silent worship. We made regular reports to the Monthly Meeting, and it felt like a remarkable confluence of the personal and the spiritual, a heartfelt listening to Divine guidance flowing through the three of us, along with somewhat practical steps as to research, teaching and the emergence of a sense of leading for this work.

For nearly twenty-five years, I had worked at an interdisciplinary, student-centered, alternative doctoral program at Union Institute and University, an accredited doctoral process focused upon adult learners who had a passion for learning and study, but needed mentoring and advising to complete a doctoral degree. I wanted to enact that model by an engagement in research that mattered, followed my passion, and connected to my Quaker spiritual community. Therefore, I called upon a Quaker Clearness Committee to assist me to become spiritually focused, centered in community and to engage in research that was connected to community and reflected the depths of my own Quaker experience. We came together monthly for two hours to engage in a deep listening process, open to the Spirit's influence, to the Divine, about my research, writing and teaching about the meanings of Quaker silent worship.

We were close; we shared intimate things about love, life, and longing. This group process, the sense of being held together, sewed one to another by the stitches of the Spirit, mattered to the process. I did homework; the two others probed, asked questions and wondered. "What do you really want to know about silence? What matters to you about worship? How do you get beneath the surface of words? Might there be a leading here? Might this entire research project be a form of ministry after all?"

The experience of worshipping together felt rich, layered, and intense. Early in this clearness process, I spoke out of the silence, searching for direction and clarity for my proposed research about Quaker silent worship. It didn't feel so much like asking for solutions to particular problems; rather, the important question was how to be open, to create together a resonant, spiritually informed container so that all of us could feel the presence of a Divine sense in our deliberations.

The emerging theme of ministry meant that the Clearness process became identified as a Quaker "leading," and the research became a way for me to express spiritual connections and the commitment to enact my emerging identity as a Quaker scholar. This research process welcomed forms of inquiry such as prayer, contemplation, *lectio divina* and uses of silence as part of the research design.

My engagement in a clearness committee process transformed social science research into forms of devotional inquiry. I was a researcher whose main business was to listen deeply. As a peculiar conduit, I allowed meanings to flow through and into me, and for some transformation to occur.

My research became more than an individualistic inquiry. I felt carried along by my committee members, transported as if in a river of inquiry, listening to them, influenced by their own searching. I felt a communal flow, one that offered a group gestalt, the role of researcher as community participant. This clearness process had implications for how I interacted with nearly fifty research participants, all of whom were members of the Religious Society of Friends. Because I expressed my scholarly role as a form of spiritually informed leadings, I could attempt to be simply present in interviews with most participants. I could invite prayer and openness to the Spirit into the research conversation.[21]

Research Seminars and Similar Workshops

For workshops modeled on the QIF Summer Research Seminars (SRS), it can be helpful to use a variety of formats for worship and worship sharing, as well as meditation, collaborative review, brainstorming, and other steps in dialogue and writing processes as these allow varied kinds of openings and have a synergistic effect.

In many of the SRS sessions, participants are sharing independent projects. The kind of discernment processes appropriate in such a context may be different from those needed in a larger collaborative project in which the nuanced methods of a Meeting for Worship for the Conduct of Business might be appropriate.

In these SRS sessions it has been helpful to start each day with a brief check in, followed by an open and unstructured Meeting for Worship of a half-hour to one hour in length. Then one participant will speak out of the silence, presenting material in whatever format is found to be most appropriate. This could be simple narrative, an illustrated presentation, or another type of presentation. It is helpful to circulate material in advance.

The follow-up dialogue and response can take different forms depending on the material and leadings of other participants. Often the deepest insights come after the opening presentation and initial discussion, when we move into worship and participants are asked to speak out of the silence. Sometimes it is most helpful to have everyone take part in a brainstorming session to generate ideas for the work. It may be useful to walk through a text with specific comments, if it is at an advanced stage and in need of fine-tuning.

It is helpful to have someone other than the presenter clerking the session to monitor the flow and suggest shifting modes as appropriate. The clerk may call for silence when appropriate—if, for example, the heat and excitement of the discussion is pulling people into a partisan debate rather than opening them to shared discernment. The clerk may invite people who have not shared responses to do so, or suggest useful structures to organize the discussion. For example, the clerk might split the session into sections focused on different aspects of the topic. Participants who have highly technical points might pursue them over lunch or on a hike.

When participants are gathered in a research seminar or workshop to collaborate on a project they are undertaking as a team—or working on themes in a common area—other methods may be helpful. Participants may be invited to reflect on a focused theme or question. Sharing then uses the method of "Quaker Dialogue" which is initiated by reading a query on the topic, e. g. "What role does nuclear energy have in addressing climate change?" or "What is the real meaning and challenge of poverty?" Then some opening silence is followed by people going around the circle, speaking out of the silence, sharing from their own experience and knowledge, and allowing some silence between each sharing.

Part of the worship process is the increased focus that brings a simplification of the vision and better understanding of issues. In wrestling with extremely involved issues, it is helpful to recall the words of one Summer Research Seminar participant (Laura Holliday): "It is not that complicated really." Simplifying is a norm for this work. We need a vision of the world that is simple enough to understand so we can live it in consent, voluntarily. Such simplifying is not meant, however, to become simplistic. The aim is to see things in the light of systemic connections that bring unity to the insight and focus to the analysis in ways that provide clarity and the simplicity that comes from seeing things as part of a whole.

Often some of the most helpful comments and responses come later over lunch, on a hike, cooking together, or in an open Meeting for Worship on a subsequent day. Socializing and social bonding play important roles in building a research community. The schedule should leave plenty of time and opportunity for these relationships. Sharing leadership is another strategy for building community. The more participants are involved in designing and leading activities, generally the more effective they are.

Physical proximity is a big help for this work. Community is important for motivation, worship, discernment, prioritizing, creativity, and identity. Time is also important. There is a cumulative effect of varied formats and proximity that results in increased synergy.

However, it is also possible for people to collaborate via email or conference call, especially if they have already developed a sense of each other and of community. In Circles of Discernment, meetings by conference call can keep the work moving forward on a Spirit-centered basis adding to what is possible simply through email exchanges.

Futures-invention—some kind of utopian thinking or visioning—is integral to the process of Spirit-led research. It continues a tradition begun by 17th and 18th Century Quakers, George Fox, Margaret Fell and John Woolman with their "openings." The workshop "Imaging a World without Weapons (IWWW)" developed by Friend Elise Boulding and Warren Ziegler, in which participants image the future in depth, can be useful either for individuals to generate research insights, hypotheses, and visions to frame their work, or for groups undertaking a shared project. The multi-day, seven-step process includes:

1) Forming values and goals that frame that possible future,

2) Warming up the imagination with memory exercises,

3) Individual inward imaging, jumping thirty years in the future with guided meditation,

4) Forming imaging teams of people with leadings toward a common future,

5) Elaborating the imaging and discerning in detail using a wide variety of methods (scenario building, inward imaging, brainstorming, drawing, role plays, macro-analysis),

6) Remembering your way back to the present, and

7) Discerning actions to work for the desired future or start living it in the present.

Central to all seven steps is the discerning practice of working out of the silence and practicing deep listening. While a full scale IWWW workshop can be especially enriching for a research seminar, more abbreviated imaging activities can also be helpful to provide space for wider vision and utopian thinking that allows people to discern important leadings.[22]

Summer Research Seminar Experience
—Laura Ward Holliday

I was introduced to the Quaker Institute of the Future (QIF) while attending the General Meeting of Friends Committee on National Legislation (FCNL) in November, 2005. I knew nothing about the organization, its mission or the emphasis on research. After my initial conversation with a founding member and further investigation, I discovered an alignment of my interest and the goals of the organization to inform public policy through research and other means with the purpose of transforming our society both domestically and globally for the benefit of the commonwealth of all life. I decided to attend the Summer Research Seminar the following year.

I arrived in Bar Harbor Maine with no topic and no idea of what would happen. We began the sessions every morning in Meeting for Worship for the Conduct of Research. Because I arrived without a topic and knew no one, every morning I would settle into the silence and wait. In the afternoon, while others were working on their topics and presentations, I sat and waited. For the next ten days, no topic was revealed. I began to wonder and ask myself, had I made a mistake in coming? Although uneasy, I had the sense I was in the right place for the right purpose. I had never approached research this way nor participated in an event with no knowledge of why I had decided to come. I just knew I had to come. As the days passed, I listened to other presenters and shared ideas and thoughts about their projects. For me, the daily communal discernment seemed to deepen.

The communication between my inner self and my Source required a stillness and space that could not be identified, but was expanding. It emanated from an unseen place, not from a place of intellect, but from an inner sense of Being. It was as if my physical ears turned inward, and I was listening as the topic arose from the silence of my soul. Finally on the eleventh day, out of the silence my topic was revealed. It was then that I could engage in classic methods of research. The process required stillness and openness to all information within me and from my Source. It required listening to unspoken words from unseen mouths and a willingness to wait, trust, and hear the whispers coming from this place. The topic that arose out of this process became a keynote speech the following year at Friends General Conference entitled "The Global Transformation of Corporations, Financial Institutions, and Government: A Quaker Approach."[23]

Way opened through both personal and communal discernment, a willingness to wait and trust that it would happen. This experience demonstrated and affirmed that out of worshipful silence, expectant waiting, stillness and a desire to hear, revelation happens. Spirit-led discernment, critical participatory research can and will produce tangible results.

Circles of Discernment
—Keith Helmuth

QIF has developed a Circles of Discernment (CoD) program that encourages Friends to undertake research using the collaborative methods and collective discernment of Quaker process. CoDs are small groups that study and research topics of critical importance for the future of humanity and the wellbeing of Earth's whole commonwealth of life. "Study" means a more systematic approach to the investigation of a topic than would be the case with casual reading. "Research" in this context means secondary research— the review and synthesis of information from existing sources. "Collaborative discernment" means the shared allocation of study, research, and reflection. It means collectively reviewing what has been learned and deciding how to frame and present this knowledge in the light of Quaker values and Spirit-led dedication to the common good.

CoDs can be formed in a variety of ways. A person with a particular interest can take the lead and create a CoD within their local Meeting or regional area. Persons at a distance who share a common interest and concern can create a CoD by utilizing email and conference calls. CoDs conducted at a distance through electronic means have been rewarding and successful. An effective coordinator is critical and a strong commitment by participants to systematic communication is essential. It is helpful if participants already know each other or if they can meet for face-to-face work during the project.

Although the method could be used for other outcomes, to date QIF-sponsored CoDs have prepared pamphlet-length manuscripts on their findings. These studies have been published as QIF Pamphlets of approximately 70 to 120 pages. QIF Pamphlets are oriented toward prompting further study among readers, and toward making a Quaker contribution to the larger public dialogue on how human communities can live in right relationship with each other and with Earth's commonwealth of life.

In developing a collective manuscript, it often works well to have different participants prepare specific sections of text. This method usually requires some re-writing by a lead editor to harmonize voice and style. Another option is to have specific sections written by individual authors as distinct essays. A third option is for a lead author to prepare a draft text that is then collectively reviewed, supplemented, revised, and edited until participants are satisfied with the outcome.

Different CoDs have utilized these different approaches. The key to making them work is to engage in open and reflective discernment, which for Quakers includes substantial periods of gathering in communal silence to see what insights may be given that help advance the work. Experience has demonstrated that this practice can be successfully conducted during conference calls. Participants in CoDs that have brought their manuscripts to completion, commonly realize that what they have created exceeds what any one participant could have accomplished by working alone.

CoDs can take various forms and durations. The key factor is for collective discernment to create a bond of guidance that draws the project into forms of knowledge that provide higher insight into the ethics of the common good and the practice of right relationship.

Reflections on *Fueling Our Future*
—Robert McGahey

I tremendously appreciated the CoD process that resulted in the QIF pamphlet, *Fueling our Future*. As a modest defender of the nuclear power option, I found a much more collegial reception than at another discussion of a minute on nuclear power where I felt only the clerk supported my right to speak. On this occasion, I was in fact asked to sit down and to stand down by others present. By contrast, discussing such a contentious topic coming out of worship or through a process of quiet, searching discernment felt much more like a search for truth, and I felt mutual respect in the group, though we ended up differing on the sustainability and health issues of this energy option. I encourage further use of the CoD process, and of sharing the method and philosophy with other groups and in other settings. Thank you all for including me in the process.

Reflections on *Beyond the Growth Dilemma*
— *Pamela Haines*

It was helpful to meet with a larger group (about half of whom were not part of the writing project) over the course of nine or ten monthly meetings to put out our ideas on the broad range of topics that were relevant to the pamphlet (*Beyond the Growth Dilemma*). I went back to the notes from the meeting that related most closely to my chapter to be reminded of topics and angles to be included.

I have to admit that one thing I don't like about group writing projects is being edited. I like to do my own editing of my work, and can often do a pretty good job, so getting pretty extensive critical feedback from four or five readers was quite challenging. But it was clear how the incorporation of many of the comments made the chapter stronger. I was enormously helped by one of the long-time QIF members reminding me that I did not have to incorporate all of everybody's suggestions. So there's an issue of discernment about what to respond to, and what to let go by. I find myself interested and puzzled by how to know which is which.

I was fascinated to be in on a discussion by three of the primary writers about what to do when the writer of one of the chapters baulked at accepting editorial suggestions. It made me realize that, while I didn't welcome it, I was ultimately teachable. I saw that those three people had a much deeper understanding than I did of how the pamphlet needed to unfold, and why this chapter in its current state was so problematic. I had been approaching it on a much more superficial level (is the writing clear, does the argument move forward coherently?), and missing a piece of the big picture. So, I feel like there was a higher level of discernment that was going on than I was aware of. I don't know if there would have been a way to invite more people into that process more deeply, or if it was just good that it was happening. Mostly, I was just appreciative of the perspective and thoughtfulness they brought to the process, and glad to get a glimpse of it.

Somewhat unexpectedly I worked with one of the two overall editors on the final chapter, and gradually took on more and more responsibility for the shape of it. I'm passionate about sending people away from a written piece like this with a sense of what they can do, and I was glad to be invited to play a role in making that happen. So her sense that she needed some help, along with my willingness to step up, made for a good partnership. The other overall editor also weighed in with some changes that made the ending stronger. This all happened over e-mail without any formal discernment process, but my sense was that we were well enough in tune with each other that this was not a problem.

Book Project: The Experience of Writing *Right Relationship*
—Geoff Garver and Keith Helmuth

Quaker Institute for the Future (QIF) initiated its Moral Economy Project in 2005 with the goal of producing a flagship book for the Institute. The writing of *Right Relationship: Building a Whole Earth Economy* was accomplished through a process that included research, working-group discussions, written contributions, and writing team discernment, all inspired by Quaker tradition and practice. This QIF-sponsored process leading to the publication of *Right Relationship* is an informative case study of the Institute's goal to promote collaborative research undertaken in the spirit of Quaker discernment.

The experience of producing *Right Relationship* invites consideration of how conviction and humility are in constant interplay in the discernment process called "Meeting for Worship for the Conduct of Research." George Fox counseled Friends to "be patterns, be examples in all countries, places, islands, nations, wherever you come; that your carriage and life may preach among all sorts of people, and to them." He said that the return for bearing this witness is that "you will come to walk cheerfully over the world, answering that of God in everyone."[24]

Implicit in this powerful exhortation is conviction. To express oneself as a pattern or an example, one must have a strong leading or conviction of the rightness of one's action or belief. Yet, the unending quest for continuing revelation also means that some degree of uncertainty underlies all such leadings. This is the inescapable check on faith—and faith is perhaps the ultimate manifestation of conviction. The inescapable uncertainty implicit in continuing revelation calls on Friends to be humble, even where convictions are strong and well tested by discernment and unity among Friends. The process leading to the publication of *Right Relationship* reflected this tension between reaching a level of conviction and having continual revelation that challenges that conviction.

Participation in the *Right Relationship* project was open: QIF placed an invitation in *Friends Journal* to solicit Friends for the project. QIF was fortunate to have a donor who generously sponsored the research and writing of the book. This support allowed QIF to convene a series of retreats to work on the book, and to provide modest stipends to authors. Participants in the project had their first gathering in January 2006 and met again in June 2006, January 2007, and June 2007. These gatherings were centered on sessions clerked in the spirit of Quaker process, with ample time for communal reflection in silence. The project convener, assisted by others, compiled and distributed background texts, draft outlines of chapters and other preparatory materials prior to the gatherings.

Over the course of these gatherings, the book evolved from brainstorming to outline to a collection of chapters authored by participants. By Fall 2007, the book writing team had assembled a complete but somewhat unorganized and uneven rough draft that was packed with a rich jumble of ideas. This draft was circulated to the QIF Board and to a broad range of Quaker and non-Quaker outside readers, including many academics. A number of these readers were then convened into a series of geographically convenient discernment group meetings to provide critique and advice.

The response from these supplementary discernment groups was highly encouraging. At one meeting a senior figure in ecological economics startled the participants by saying "this manuscript should be shredded," but then quickly explained that he had said this because the book's subject and our approach to it is so important that we should start again and make it better. A professor of law told us she had been looking for a book like this to use in her course on international environmental governance but had not found one. She said, "If you get this book published, I will use it." In commenting on some overly technical parts of the manuscript, one participant suggested that in rewriting we think of our audience as undergraduate college students.

At this point in the process, it was clear we did not yet have a coherent book, but with this kind of helpful advice and encouragement, a smaller team of core authors began rewriting the manuscript in the winter of 2007-2008.

Around this time, the project convener found a publisher interested in shepherding the book to publication. This was an exciting development. The publisher's editor then became, in effect, part of the discernment process and offered a key suggestion that changed the structure of the book. The original manuscript was founded on the core concept of right relationship and focused on five basic questions about the economy, which it posed in the first chapter and then went on in successive chapters to build the case for the "whole Earth economy" that our answers envisioned. Our editor told us that if we write one chapter on each question from the ethical perspective of right relationship, that is the book she wanted to publish. And she said to aim for 200 pages with a final draft due by July 1, 2008.

In January 2008, three of the principle authors met for a weekend to review the work that had been done so far and set up the process for recomposing the book according to this new plan in collaboration with three additional authors. We had a common sense that something rather extraordinary had quietly happened in working together. Having started our session with a period of centering silence, a smoothness of collaboration unfolded in which a unity of spirit and vision was clearly evident. Individual authors took primary responsibility for drafting specific chapters. The other authors also contributed to each chapter. During the next nine months, repeated circulation of all chapters to all authors continued. Two authors in the same location worked intensively to coordinate editing and bring the final manuscript to completion. The result was a book of truly collaborative authorship.

The introduction of a publisher and a thoughtfully engaged editor in the book's gestation intensified the book-writing project, primarily by introducing the publisher's criteria and by imposing a fairly tight deadline for completing the manuscript. The publishers' requirements tested the group's process of coming to unity and maintaining a sense of the meeting with the QIF Board of Trustees. For example, the project convener wanted an understanding that he would make time-sensitive decisions unilaterally if necessary. The book's authors were not entirely at ease with how this approach would impact collaborative discernment and a sense of unity regarding the book's content. When, in fact, the authors did not reach unity on the final content of the manuscript that was submitted to the publisher, a suggestion was made to consider that publication of the book was a somewhat artificial milestone in regard to ongoing reflection on the matters discussed in the book, and not the end of that process. This proposal was not sufficient to overcome concerns, and one participant withdrew as a named contributing author.

The QIF Board of Trustees was able unite on a statement about the Moral Economy Project and the book that includes the following:

The Quaker Institute for the Future sponsored, and was pleased to support, the team project that resulted in this volume. QIF saw the book's special merit in clearly identifying the great crisis of our time and offering principles to be applied in response. Its findings contribute to an ongoing dialogue process that needs to engage public policy professionals as well as a broad range of citizens. While the views expressed in this book are those of the authors, the work is endorsed by QIF's board of trustees, which is united in its sense of urgency about advancing dialogue toward prompt critical action needed to rectify the growing incoherence between the human economy and the integrity of earth's ecological and social systems.[25]

Right Relationship was published in January 2009 and has gone into a second printing, with over 7,500 copies sold. It has been used in a number of courses in colleges and universities. It has been distributed in Europe and Africa. It has been used in an adult education English class in Poland. It is being translated into Chinese and published by General Publishing in Beijing.

The experience with *Right Relationship* also invites a consideration of how Friends involved in discernment for the conduct of research might deal with questions of monetary compensation. The generous budget that supported the development of *Right Relationship* made it possible to reimburse travel expenses for retreats and other meetings; administrative expenses; stipends for the project convener, research assistance, and contributing authors; and expenses for a post-publication symposium. For other QIF projects and the QIF Board of Trustees, the supporting budget has been limited and participation totally voluntary in the facilitation of CoDs, and the production and distribution of QIF pamphlets.

Lessons from a Quaker Approach to Research

Several lessons emerge from the QIF experience in communal discernment. They lead to the following questions for research and writing teams seeking to follow a Spirit-led Quaker process.

- Will this work have a distinct single author, sections written by different individual authors, or a coherent manuscript created from all participants' written contributions?

- To what extent do the likely participants in the process share values, visions, skills, personality traits, bodies of knowledge and other resources which complement each other in ways that might further and/or impede the discernment process? Are there voices missing that should be sought out?

- Are those participating ready to commit to following through on the project as a shared discernment process? What might be done to strengthen the resources or commitments of the group?

- To what extent might some or all of the issues at stake be better addressed by an alternative approach? To what extent are they purely technical questions that call for expertise within a narrow and standard body of disciplinary knowledge that could be provided more efficiently by a single professional?

- To what extent are they issues that call for reaching a political agreement between conflicting constituencies who might have a strong preference for using a more formal procedure of decision making like Robert's Rules of Order because it is legitimated by legal statutes governing the organizations involved or because it presupposes a strictly secular frame of reference that might be more acceptable to the parties involved?

- How will the team manage any timelines or deadlines, which may vary in terms of strictness? For example, what situations might warrant extending a deadline, taking into account the nature of the deadline?

- What matters regarding decision making should the team decide on early in the research and writing process, e.g. might the team seek unity early on in regard to having one or more team members making time-sensitive decisions where necessary?
- How will the team decide that the research and writing project is finished, taking into consideration that every written work represents but one point in an ongoing evolution of thought, discernment, and intellectual development? How does this decision and others in the research and writing process draw on the balance between leadings of faith and conviction, and openness to continuing revelation?
- How might additional funding enhance or complicate the research and writing project? Is it worth seeking funding in order to enhance the desired outcomes of the project, by supporting authors, convening multi-day retreats or working sessions, facilitating involvement of a wider set of reviewers, or allowing for broader distribution of publications?

Summary

In the ten years since QIF was founded, experiments have been tried with a variety of methods for practicing meetings for worship for the conduct of research. These have included Clearness Committees, Circles of Discernment, and a book writing group. They have also included a series of summer research seminars whose participants have experimented with variations on these and other Quaker-inspired practices such as Threshing Sessions, Quaker Dialogue, and "Imaging a World Without Weapons" workshops.

These experiments have produced a growing body of research and publications that reflect Quaker values in the substantive analyses they offer and also exemplify Quaker practices of communal discernment in the ways in which the research has been pursued. From the time of their first gatherings as a Religious Society of Friends of the Truth, Quakers have held that the life of the Spirit is not something to be practiced only at some segregated time and place such as a "steeple house" on a Sunday morning.

Instead, they have held that all of life is a sacrament and that all of life should be viewed as an opportunity to attend to leadings of the Divine and guide actions in its Light. For the participants in Quaker

Institute for the Future activities described here, this conviction has been strongly reinforced as it applies to activities of research. Research calls for keen observation and critical thinking. But it need not call for laying aside our ability to also listen to the leadings of the Spirit as we collaborate in seeking to discern the truth.

Quakers are not the only or even the first to use methods of communal discernment. And the particular practices they have developed can and should be enriched by dialogue with other traditions of Spirit-led discernment, as well as secular practices of collaborating in the pursuit of truth. It could be very fruitful to pursue such comparative dialogue in considering in what practices might be better at promoting successful collaboration that generates insight, knowledge, useful policy and action. In what ways, and to what extent, might proper use of different practices provide confidence in the trustworthiness or truth of the conclusions reached?

As part of such dialogue, it would be helpful, further, to consider, the meaning of key ideas like the Quaker notion of "continuing revelation" and how such revelation might be understood in the context of bodies of knowledge and policy that are intentionally secular and grounded in separations of faith and reason, and of religion and science. It is important to wrestle with these questions because the answers may help to revise and improve the practice. Sharing the methods and the results of the work with the larger world beyond the Quaker community is also important. The next chapter explores the experience of traditions of other communities using practices of communal discernment or collaborative research to arrive at the knowledge that governs their lives.

CHAPTER III
Comparing Traditions of Communal Discernment

Communal discernment is practiced in a wide variety of ways by communities that seek to base their beliefs and actions on a collective sense of identity. The use of collaborative processes complement, correct, or otherwise improve on the efforts of individuals, but even more significantly foster a sense of engagement and commitment. Communal discernment creates the shared ownership of the understandings, decisions, and actions that are required for authentic community life. There is much to be learned and shared by comparing the experiences of different traditions.

This chapter surveys the experience of some different traditions, highlights some of their features, and notes where further dialogue between Quakers and other traditions might be fruitful. The first section begins with an extended comparison with some of the processes of spiritually based communal discernment found in indigenous cultures of North America. The role of egalitarian circles in which people speak out of the silence in Spirit-led ways provides important and illuminating analogs to Quaker practices of communal discernment. Other traditions of communal discernment and collaborative research are briefly described that provide similarities and contrasts with Quaker practices in ways that may also be illuminating. These comparisons are made assuming that each practice—including Quaker practice—has strengths as well as limitations and that dialogues amongst them may help each learn from the other, enriching and strengthening each tradition.

In closing this chapter, special attention is paid to traditional forms of academic research as well as more recent developments in community-based and participatory research in order to consider not only comparisons to Quaker practices, but also the practical challenges and opportunities in combining them in research work.

Collective Discernment in Indigenous Cultures
—Keith Helmuth

When 18th Century American Quaker, John Woolman, traveled to the north-central area of the Pennsylvania territory to see what he could learn from indigenous people, he opened up a quality of relationship between the Society of Friends and North America's indigenous communities that has continued to the present time. Woolman's intent in making this journey was clearly a kind of research. He wrote the following about the journey:

> *Love was the first motion, and then a concern arose to spend some time with the Indians, that I might feel and understand their life and the spirit they live in, haply I might receive some instruction from them, or they be in any degree helped forward by my following the leadings of Truth amongst them.*[26]

His narrative reveals that these meetings resulted in something very like cross-cultural communal discernment:

> *So near evening I was at their meeting, where the pure gospel love was felt, to the tendering some of our hearts. And the interpreters, endeavouring to acquaint the people with what I said, in short sentences, found some difficulty, as none of them were quite perfect in English and Delaware tongue. So they helped one another and we laboured along, divine love attending. And afterwards feeling my mind covered with the spirit of prayer, I told the interpreters that I found it in my heart to pray. ... And expressed my willingness for them to omit interpreting; so our meeting ended with a degree of divine love. And before the people went out I observed Papunehang* [a community elder] *spoke to one of the interpreters, and I was afterwards told that he said in substance as follows: "I love to hear where words come from."*[27]

A striking similarity between Quaker process and Aboriginal communal practice emerges in the way collective discernment is used to construct knowledge and provide guidance. The process of open communal search, reflective communication, and its form of decision-making steered the Society of Friends into a tribal-like form of spiritual and social life. The Society of Friends, along with all else it might be, is a living version of our ancient, durable and essential communal heritage. This is the first and most basic level of affinity between Quaker practice and Aboriginal traditions. Both traditions approach the construction of knowledge and the emergence of guidance through communal process.

In the late 1970's, Quakers in Atlantic Canada had a particular experience of exploring this affinity when Noel Knockwood, a Nova Scotia Mi'kmag elder, was invited to participate in our annual retreat with a view to sharing spiritual insights and political concerns. He later expressed his astonishment that an organization of non-indigenous people were practicing a form of spiritual gathering and methods of communication and decision making that were so close in feeling and process to the traditions of his culture. Friends have been invited to join First Nation communities in certain of their gatherings and ceremonial practices and persons from First Nation communities have become regular attenders at Quaker gatherings in Canada. The often-used circle seating, waiting in silence, and speaking out of the silence around the circle draw the two traditions into a shared form and a common sense of presence.

All this may seem remote from the formal practice of research aimed at the construction of knowledge, but, from a Quaker point of view, this tradition of practice in both Quaker and First Nation communities is the cultural source for a new approach to research that can better create knowledge and guide decision-making that serves the common good. There is a need for "research on research" in order to determine what, in fact, is important for advancing the common good and the well-being of Earth's ecosystems. In this area of value-based decision-making, collaborative discernment, under the guidance of dedication to the common good, can be applied to positive effect.

Traditional First Nation communities of North America are not alone in having preserved the practice of communal discernment in the construction of knowledge. Indigenous communities worldwide still operate within the practice of communal discernment as a way of coming to know what they know. Over and over again it has been the case that this knowledge is accurate, vital, and critical to the wellbeing of communities and land.

Likewise, Quakers are not alone in the deliberate practice of collective discernment and collaborative decision-making on matters of policy and action. This ancient, spiritual-social form, rediscovered by the Religious Society of Friends, has now become acknowledged as a preferred form of organizational and social conduct in many situations. Advancing this process as a new approach to research further develops the likelihood that research will, in turn, advance the common good.

An excellent example of how communal discernment works in a Native American community is given in the novel, *The Man Who Killed the Deer*, by Frank Waters.[28] In this novel he tells the story of a young Hopi man, Martiniano, who killed a deer out of season in an area of forestland controlled by the Federal government, and without ceremonially granted pueblo permission. He was arrested and, in the process, beaten by police. A pueblo council meeting is called. The meeting begins with the burning of pungent cedar. Tobacco is handed round and smoked with deliberation.

> *Now the silence, impregnated with smoke, seemed thicker, heavier ... As each man threw away his stub, he leaned back against the wall and drew the blanket up around his head. They might have been settling in for a long sleep. Then suddenly the talk. Slow, measured, polite and wary.*[29]

The Governor of the pueblo introduces the matter at hand. Then both of Martiniano's companions tell of the incident. Silence. Then Martiniano tells his story. Silence. Finally, the talk begins again, starting from the simple facts: A deer killed out of season; a Hopi man arrested; a fine to pay. The talk then develops in ever widening circles around these facts. The man was born in the pueblo. Although he has failed to respect his ceremonial responsibilities, and has been repeatedly warned, yet no person is single and alone. The pueblo is involved. Still more questions are raised. It is a matter of the land. The mountain where the deer was killed is Hopi land, land that was never relinquished nor legally obtained by the government. What will come of this continued usurpation of sacred Hopi land? This matter of sacred land involves the ceremonies and the whole life of the pueblo. There is the matter of the deer, killed without its ceremonially granted permission, without proper ritual awareness of the way all that is living flows together. The issues widen, the considerations deepen, centering down into the essential facts of Hopi life and consciousness. The speaking stops; silence continues the communication. Frank Waters then articulates the silence.

> *Silence spoke, and it spoke loudest of all ... There is no such thing as a simple thing ... Nothing is simple and alone. We are not separate and alone. The breathing mountains, the living stones, each blade of grass, the clouds, the rain, each star, the beasts, the birds and invisible spirits of the air—we are all one, and indivisible. Nothing that any of us does but affects us all.*

A council meeting is a strange thing. The fire crackles. The candle gutters. And the old men sit stolidly on their benches around the walls. When a man speaks they do not interrupt. They lower their swathed heads or half close their eyes so as not to encourage or embarrass him with a look. And when the guttural Indian voice finally stops there is silence. A silence so heavy and profound that it squashes the kernel of truth out of his words, and leaves the meaningless husks mercilessly exposed. And still not a man speaks. Each wait courteously for another. And the silence grows around the walls, handed from one to another, until all silence is one silence and the silence has the meaning of all. So the individuals vanish. It is all one heart. It is the soul of the tribe. A soul that is linked by that other silence with all the tribal councils which have sat here in the memory of man.

A council meeting is one-half talk and one-half silence. The silence has more weight, more meanings, more intonations than talk. It is angry, impatient, cheerful, but masked by calmness, patience, dignity. Thus the members move evenly together.[30]

The role of group silence—communal meditation—is one of the keys to the collective discernment that underlies this kind of collaborative research. The construction of knowledge and the emergence of guidance are fully reflected in the last line quoted above: "Thus the members move evenly together." Quakers know from the experience of their practice how this "moving evenly together" happens and what it means for effectively advancing policy and action on collective concerns.

The cross-cultural heritage of this approach to the construction of knowledge is anchored in a human universal—the spiritual dimension of communal experience. When research is conducted using this approach, the knowledge gained carries a kind of wisdom about right relationship and the common good that is critically important for the well-being of human communities and the health of Earth's ecosystems. Over three and a half centuries of Quaker experience with collective discernment and the immemorial communal heritage of indigenous peoples bears witness to this way of knowing.

Spiritually Framed Traditions of Communal Discernment

There are many other practices of communal discernment that grow out of spiritual traditions. While these often frame their understandings of the nature of reality and knowledge in dramatically different ways, they may share paradigmatic experiences and core elements that could provide a common ground for very productive dialogue.

Shared paradigmatic experiences may include ways in which an encounter with silence or music may be enriched when in community with others. The specific forms of these encounters may differ significantly within Jewish, Christian, Moslem, Bah'ai, Hindu, Buddhist, and other traditions. They may combine outward silence with structured inward mantras, prayers or other forms of focused thought. They may try to achieve a form of inward silence as well as outward quiet. They may use outward motion in dance or vocalization in song that serve meditative functions and distance themselves from the smaller, petty, grasping, fearing, and desiring self of the everyday. Shared elements of practices in these traditions may include the use of meditation to prepare for entering more fully and openly into a discernment process and ways of allowing for continuing revelation.

Traditions of faith and spirituality may share egalitarian approaches to discernment. This can spring from the view that there is one God through which all believers share a common bond as in the Abrahamic faiths. Or, as in some traditions of Buddhism, it may spring from a view that affirms the interconnectedness of all beings and the ways compassion calls for a kind of sharing that has profoundly egalitarian implications. There may be limitations based on who counts as a believer and who is viewed as competent to speak according to gender, age, or formal education. But within the community so defined, there may be practices of egalitarian debate and discussion that enhance the discernment process by seeking to keep the dialogue open.

Although this type of comparison is beyond the scope of this pamphlet, it might prove useful to compare traditions of rabbinical dialogue in the Jewish tradition, the processes of exchange surrounding fatwas in communities of Moslem scholars in the Sunni and Shi'ite traditions, practices of conflict resolution in the Socially Engaged tradition of Buddhism, and practices of discernment in Jesuit communities. This kind of dialogue between traditions is exemplified by Jesuit Michael Sheeran's detailed ethnographic study of Quaker practice in the Philadelphia region. The result was an especially illuminating analysis that many Quakers themselves find helpful in articulating and improving their practice.

Where Quakers had been often tempted to focus on the ways they share common experiences but not press each other to define their terms, Sheeran's work brought a kind of philosophical rigor to the study of Quaker ways, which included experiences of finding unity, of having "covered meetings," and of encountering the Divine. This way in which someone trained in one tradition, the Jesuit, may bring language, experience, interests, and skills to the study of another, exemplifies just the sort of interfaith dialogue that could prove very rewarding for all concerned.

Extending Quaker Practice to Secular Settings

What about going beyond the inter-faith context? What might be learned through comparative dialogue about communal discernment as practiced in explicitly secular, non-religious contexts?

People coming out of religious traditions who work in secular contexts have brought elements of their faith-based discernment practices into the secular context with very significant success. Often this involves translating the concepts and practices of religious traditions into secular terms in a way that preserves their functional aspects without requiring secular practitioners to adopt specific spiritual or theological commitments.

An excellent example of translation of Quaker idiomatic language to secular terms is found in the way Leonard Joy discusses the essentials of Quaker practices in language appropriate for secular settings.[31] The practices includes:

- grounding of all participants in the desire for the common good,
- preparing factual and analytical material prior to the meeting,
- ensuring that all voices are heard and listened to,
- respect for all affected by the decision-making process,
- sensitivity to interdependence—open systems thinking,

- speaking out of the silence,
- addressing the clerk not one another,
- speaking simply and not repeating what has already been offered,
- speaking one's own truth without advocating that all should act on it,
- a commitment to express reservations,
- being authentic with the expression of feeling without using emotion to sway others,
- distinguishing "threshing" meetings from meetings for decision making,
- clerk offering syntheses of the "sense of the meeting" that are modified until there is unity,
- making decisions not by majority vote, nor by consensus, but by unity.

Quakers who were part of the Movement for a New Society in the Philadelphia region in the 1970s developed a process that was widely and successfully adopted by affinity groups and other social movement organizations in secular contexts.[32] They focused on procedural elements such as making sure that all participants had opportunities to share in a respectful dialogue, checking for unanimity rather than taking votes, and allowing people to stand in the way or be asked to stand aside. They avoided spiritually loaded terminology, such as, references to the Inward Light.

Similar strategies have been used by Quakers in a wide variety of other secular settings.[33] In many of these contexts, it can be challenging because the groups involved have very diverse values and beliefs. Among Quakers, the process of communal discernment is facilitated by agreement on core values and testimonies about peace, equality, simplicity, Earth stewardship and, most importantly, their conviction that there is "that of God in everyone."

One way to provide common ground in a secular setting is to ask participants to agree to pursue their dialogue with the aim of seeking the "common good." This asks them to take into account the good of all the participants and constituencies represented, and introduces an egalitarian element that serves a similar function to that of there being that of God in all. The "common good" presupposes that there is a vision of the future that the participants share. It invites them to seek agreements that rise above the level of political manuvering to reach genuine, sustainable, common understandings analogous to the Quaker notion of "unity."

Other terms that could play a similar role to "the common good" in this context might be "sustainable living" or "common future," in so far as they imply the basis for a shared understanding that is affirmed by the participants. One common feature is that they invite participants to shift pronouns and the identities they express, to move from talking exclusively about what "I think and want" with regard to "my personal interests" to talking inclusively about what "we think and want" with regard to "our shared concerns."

Experience of Quaker Process in a Secular Setting

—Leonard Joy

One of my most memorable experiences in the use of Quaker practice for decision making in secular contexts was in a former Soviet country after the fall of the Soviet Union. I was charged to support a team created to manage a process for the redesign of the public sector. I acted as chairman-moderator of their formative meetings for planning their work. In that role, I chose to act as would a clerk in a Quaker meeting. The team included former government officials and a member of the secret police.

The culture of decision-making to which they were accustomed required them, each in turn, to endorse proposals sent from higher authority. Released from this constraint on their self-expression, they could not contain themselves from argument, shouting above one another, and interrupting for their voice to be heard.

They acknowledged that this was not productive and accepted my clerking authority, which now required them to open their meetings with silent worship. Of course, I did not call it that. I asked them to center themselves in their role in search for the greater good. I saw that as another way of responding to the awareness of God in themselves. And what is that if not worship?

I also asked them to speak only when acknowledged by the clerk— which, of course, I called "the chair." I further asked them not to present arguments against one another but to each contribute what they understood that was relevant to the decision. I emphasized that we should use the ego to serve the task and not the task to serve the ego.

I further explained the aim of coming to unity and the value of that in securing ownership of the outcome. In my role as chairman I gave periodic reports of what seemed to be agreed, what seemed to need further resolution, and what I sensed that this would take, inviting contributions on these matters.

We made small posters and stuck them on the wall—prompts to remind us of what was now required of us. They were referred to readily when there was departure from our new aspired-to norms. Indeed, the value of the new practice was readily seen and it became adopted with pride. The team members set out to spread this culture in the meetings they were calling in the different branches of the administration in the process of governance redesign.

****

Another comparison between faith-based and secular processes of discernment can run in the opposite direction. Processes of collaborative research and discernment in secular contexts rooted in scientific understanding can also inform or improve on the faith-framed ones. Secular traditions of natural science and social research have developed a host of quantitative and qualitative methods for observation, experiment, data analysis, and theory formation that have proved extremely powerful. They can be incorporated into any systematic approach to describing, explaining and dealing with the world. To cite just a few examples, professional methods of ethnographic interviewing, double-blind experiments, regression analysis, and mathematical modeling all have important uses. When a community of researchers is investigating questions about the food system in Bolivia or the life cycles of malaria protozoa, they can incorporate these methods as appropriate. They are analogous to the third stage of Quaker process, the gathering of points of view and the attempt to synthesize them.

The Quaker approach to research should not be thought of as an alternative to the methods of modern science. Instead, it should be understood as a proposal to enrich it, using open and inclusive dialogue in communal discernment.

Two features characterize modern scientific research in its search for knowledge: 1) appealing to public, shared, repeatable experience and 2) avoiding biases in which irrelevant prejudices exercise power over decisions about what to believe. In place of esoteric doctrine, revelation through a sacred text like the Bible, or appeal to authority like Aristotle, modern science, as it emerged with Galileo and Newton, sought descriptions of the world grounded in the results of experiments and observations that could be shared and repeated by any competent observer.

Science works by consensus rather than votes, so the practices of science are closer to those of Quaker practice than Robert's Rules of Order. If opinion is split on a theory because researchers get different results or have different interpretations of them, they work to refine their procedures and share them until they all get similar, repeatable results. They analyze their interpretations until they agree and form a community that shares a common paradigm for practicing what Kuhn refers to as "normal science." If they have trouble finding such procedures or arriving at accord in their interpretations, they may get frustrated and angry. Animosity may even result. But in their role as scientists, they remain obliged to keep looking until some way opens for them to find such procedures or interpretations. They do not pull out their guns to start shooting, and they do not call the question, take a vote, and let the 51 percent determine what is true.

Communal discernment, of which the Quaker process is an example, is similar to and compatible with the tradition of mainstream science. Both operate on the conviction that there is an objective truth independent of any individual and that with patience, such truth can be sought and found. Once found, truth can compel conviction by an appeal to rational consent that makes violent threat and authority unnecessary. In some paradigms, such as classical physics, objectivity has been formulated in terms of laws that are universal, precise, and unchanging. These laws are objective in the sense that they are invariant from one observer to another. Other paradigms, such as Darwin's evolution and Mendel's genetics, follow probabilistic laws. They have both used patterns in natural history that have highly contingent origins but emerge as explanatory structures. For example, the bilaterally symmetric structure of the line of animals from which we descend differs from other fascinating creatures, the remains of which have been found in the Burgess shale formations.[34]

The broader notion of objectivity that animates science and modern academic disciplines is the notion of a truth that is objective, not by virtue of being invariant for every observer through every perspective, but rather by virtue of being inclusive of all the variations and putting them in the larger context of the whole story of which they are a part. Objectivity in this broader and richer sense aims at impartiality or completeness that can include any Newtonian-like laws that may be discovered. Objectivity also includes the rich contingent patterns of reality that describe the shape of the continents, the descendance of animals on them, the development of social systems, and the nuances of cultural communities.

Modern academics have developed important and very useful procedures for assessing the merits of research aiming at truth. These include anonymous and blind review of communications shared in ongoing research.[35] Other important procedures include those used in certifying the skills of people who gather facts in the field or the lab through training programs that assure they have learned to observe things in the frame of their discipline and to set aside their prejudices and biases.

Such academic procedures for discerning the truth provide a rich resource of methods for Quakers and practitioners of research in other traditions of communal discernment. In principle it would seem that there is no reason why the reverse might not also, at least sometimes, be the case. For instance, when secular academic researchers use anonymous and blind review, the communication process by its very nature tends to exclude the flow of information and assumptions that are tacit and concern presuppositions that may embody new and important ideas that challenge orthodoxy but at first look like simply wrong-headed or naïve ideas. In contexts where disputes about the merits of a submission arise, it could at times be helpful to convene a meeting in which researchers practice collective sharing, reflection, and dialogue methods used in communal discernment traditions. Such conferences may allow members of the research community to bring tacit perceptions and presuppositions to the level of explicitness that allows them to write articles that communicate clearly and pass the test of anonymous and blind review.

Meetings using such interpersonal, collaborative methods can be practiced in a wide variety of settings in the research process. Academic conferences provide one example. Research teams at an institute might make use of them as part of their grant writing process. Doctoral candidates might make use of them in developing their proposals and assessing work in progress. Researchers who would eschew any reference to the "Inward Light" and would insist on a secular approach to their work could make use of silence, taking turns speaking out of the silence, listening deeply, and being open to the views of others in friendly, if not loving ways. They might be open to new framing questions that invite reflection, and continuing new insights.

Academic researchers might use such meetings to facilitate their understanding and assimilation of knowledge that comes from beyond the academic community. For example, biologists researching fisheries may find that there is a host of local knowledge that people working the waters for generations can share. Ecologists studying forests may find that native communities are a rich resource of knowledge. Such local and indigenous forms of knowledge typically operate from very different interests, employ different styles and methods of perceiving the world, and embody very different general assumptions about how the world works and how it can best be known.

Useful encounters would involve cross-paradigm and cross-cultural collaboration that may demand very careful, open minded, patient listening and sharing in interpersonal meetings and community gatherings. In recent years efforts to develop methods for science that draw on such knowledge have become an important focus of experimentation in many fields and significant priorities for institutions such as the National Science Foundation.[35]

Quaker Process in a Biology Graduate Program
—Judy Lumb

When I was a faculty member in a graduate program, our students began complaining about specific requirements, as if they were only trying to get the degree, but had no concept of their future lives as research scientists, or even of the process of research. So I developed an orientation program for new students on creativity and research based on my experiences as a Quaker. The three sessions were 1) logic, what constitutes proof; 2) honest and clear use of language, how it can be deceiving as in propaganda; and 3) intuition, how an intuitive flash can lead to important discoveries, "what if the basis of cancer were the reactivation of genes used for embryonic development where continuous growth of cells is needed?"

I had taken the "Imaging a World Without Weapons" workshop at Friends General Conference (FGC), so I used those visioning exercises in the intuition section. I had two purposes. I wanted the students to have a vision of themselves as a scientist, to have an idea of their future. And, I wanted to awaken their intuition, so they could trust in its relevance to their scientific work.

After a lecture on intuition and its role in scientific research, I explained that I was going to demonstrate. I darkened the room and did a guided meditation, placing them 30 years in the future as a research scientist. As in the FGC workshop, I elicited detailed images. At the end of the meditation, I put them in small groups of about five each with queries about their experience in the guided meditation and any other intuitive experiences. I gave them rules as would be used in a Quaker Dialogue. They should speak in turns, not interrupting each other, and leave some space before the next speaking to give time for reflection on what was said. No one should speak a second time until everyone has spoken.

The room was buzzing with intense, but quiet, conversation. Several expressed their appreciation and amazement at that experience afterwards. Those students seemed to have a clearer idea of themselves as future scientists. They were more confident and less resistant to the graduate requirements than those from the previous year.

****

In dealing with "wicked problems" like managing water resources in the American West, researchers and community members have developed a tradition of "collaborative rationality" for dealing with complexity that provides another important set of relevant, helpful practices. As Judith Innes and David Booher have noted:

> ... many scholars and professionals have become disillusioned about the adequacy of the traditional model for advising officials in contemporary times and frustrated with rigid, one-size-fits-all bureaucratic procedures and public officials so buffeted by multiple constituencies

that they are unable or unwilling to take strong action. Many professionals, in our observation, have begun to search for new ways of thinking, gathering evidence, and deciding, turning their attention particularly to communication and deliberation among citizens, stakeholders, and professionals. Collaborative practices today are spreading and evolving along with theory, as the potential players learn more about how they work and how they can improve the outcomes of planning and policy.[37]

The sharing of methods between mainstream academic research and communal discernment traditions raises some interesting and important philosophical questions. These methods come out of different traditions and frameworks for understanding the ultimate nature of reality and how it is revealed. How can their differing presuppositions be coherently integrated or navigated as researchers move, for example, from the context of an annual conference of a professional society, to meetings with a community of subsistence farmers, to a gathering of Friends? The aim of the next chapter is to articulate some of the philosophical questions that arise in this context and explore how they might best be addressed.

CHAPTER **IV**
Philosophical Issues Concerning Communal Discernment

Chapter III ended with some practical suggestions as to how Quaker practices of communal discernment might be included in mainstream secular research practices. But the idea that such inclusion and integration would be useful assumes that the Quaker practices are themselves legitimate ways of generating knowledge and that their assumptions about the nature of reality are compatible with those of secular research. This chapter is intended to advance an open dialogue about whether these epistemological and metaphysical assumptions implicit in Quaker process might be warranted in research contexts. To what extent can they be translated into terms that communicate effectively with non-Quakers? To what extent might researchers who are not Friends find some version of the Quaker approach to research and its assumptions plausible or warranted and be led to adopt them?

Detailed consideration of these questions is not necessary to make successful use of the methods and some readers may find the sometimes abstract and challenging reflections of this chapter less helpful than other material in this pamphlet and be led to pass over it. It is hoped however that the material in this chapter will be useful for readers who find philosophical perplexities intriguing and/or who find that philosophical reflection can enhance the insight with which they practice research and experience discoveries.

The chapter proceeds from points of most likely common ground and agreement about the epistemological issues, about the "theory of knowledge" and ways of justifying beliefs, to potentially more controversial epistemological claims. The approach to metaphysical questions, questions about the nature of reality, similarly starts with points of most likely common ground and then moves to points of greater controversy.

Communal Discernment as a Method of Discovery

Philosophers of science make a distinction between the "logic and context of discovery" and the "logic and context of justification." The first context applies when people are simply trying to come up with new ideas or hypotheses. The second applies when they are trying to decide whether the new ideas are good ones, whether they advance our understanding of the world, and whether we are justified in actually believing them.

In the context of discovery, scientists can use all kinds of research methods to discover new phenomena and come up with new theories to explain them. Even methods that seem random or outright wacky might be useful if they generate new ideas that help people "think outside the box" or be creative in their formulations of hypotheses. Very successful scientists who came up with what are now quite respectable theories sometimes drew on sources for their ideas that included things like numerology (Kepler's third law of planetary motion), dreams (Kekulé's discovery of the benzene ring), or observations of behavior in singles bars (John Nash's theory of equilibria in game theory). Collaborative approaches in the vein of communal discernment developed by Quakers may prove productive. Researchers may get along well, empower each other, and generate lots of new ideas. In the context of discovery, that's fine. Anything goes.

Communal Discernment as a Method of Justification

But what about when they try to test their ideas and determine whether they are useful, trustworthy or in some sense true? Then it would seem that a different and more rigorous set of criteria apply—criteria that should be normative for all researchers, regardless of their creed or subculture. To what extent might the Quaker process of communal discernment offer, not just a cool though unusual way to generate ideas or get teams to work together, but a way of testing and justifying them, a way of doing so that would have a normative force, a power to persuade all those involved in the research, whether Quaker or not?

One important flaw that descriptions and explanations of the world can have is partiality. They can leave out important parts of the story. In this regard, Quaker process arguably provides one useful way to increase the impartiality of our understanding to the extent that it is interpreted as calling for an inclusive approach that excludes no one from the process of research. Quaker process calls for all disciplines, as well as all people, to be involved in the process of investigating issues and arriving at a justification of conclusions—arriving at clearness. It therefore implies a call for interdisciplinary research that allows, for example, the specific criteria and methods of ethnographers, economists and others to be employed, but placed in a larger context, a larger holistic understanding. For example, the larger context employing Quaker process could be used when we try to understand issues around poverty or environmental degradation.

In doing interdisciplinary research, a key part of the process is that of defining the problem under investigation.[38] What "poverty" means, for example, depends, in part, on the discipline from which it is approached. Generally speaking, if a discipline is something like a community of researchers who share a common paradigm, practicing what Thomas Kuhn would call "normal science," then the problems and puzzles the discipline normally studies are ones that are generated by the discipline's own paradigm, by the values, worldviews, and metaphysical presuppositions that frame it, and by the particular puzzles that have arisen in trying to apply its central theoretical ideas.[39]

In the context of focused work within a narrowly structured discipline, the criteria and methods accepted by the scientific community in that discipline typically provide relatively clear and routine ways of pursuing and assessing knowledge. In such cases, competent individuals can simply apply those methods and criteria in their individual work or in collaboration with others without any strong need for raising questions about the theoretical framework, the values, the observation techniques or other research methods. The communal discernment about those sorts of things has already been done; the community has arrived at a paradigm and this "normal science" can proceed apace.

However, when paradigms encounter anomalies and approach contexts in which it becomes necessary to practice what Kuhn referred to as "revolutionary science" or when the problems being addressed involve multiple paradigms because they are inter-disciplinary in nature, then research can no longer be pursued in routine ways. Increasingly more of the vital research problems are of precisely this sort. This is often true of research questions of relatively limited scope that involve multiple disciplines and large, complex problems like poverty, health care, climate change, HIV- AIDS, environmental degradation, endangered species, and hunger.

These research problems can only be defined and studied by drawing on multiple disciplines and multiple forms of knowledge that are not restricted to the academic world. To characterize these issues as "problems" is already to imply that they should be solved. Yet it may be more productive to think of some of these kinds of issues as ongoing "concerns" that we try to address without assuming a technological perspective in which some expertise will provide a lasting solution.

How do concerns under investigation get framed and defined in an interdisciplinary context, when more than one paradigm is being employed and all are subordinated to some larger project? Often it is by political mandate. Leaders in government or donor agencies have their own agenda for political reasons. In order to direct research in the general area of their perceived interest, they typically come up with some phrase that is not a part of any theoretical constructs within the disciplinary paradigms related to their interests. In that sense, the term is "extra-paradigmatic."

Something like this happened in 1948 when the Truman administration introduced a contrast between the "under-developed nations" and the "developed world" to frame the social and economic issues at stake in the post-World War II context of decolonization and the spread of Communism. At regional or local levels, research problems involving multiple disciplines may get defined through political processes within state or city governments (or colleges and universities) that frame new opportunities in a sector of an economy, concerns about a watershed, or challenges facing a minority group.

In these large interdisciplinary studies, there might be a place for Quaker process in the discerning of which problems should be studied. How should "poverty" or "environmental degradation" be defined and framed? It is an important task to first define the questions at issue in order to do research on them.

When government institutions and related political processes frame an issue, partisan interests and the exercise of power may play a large role. The framework that creates it may not allow the emergence of a coherent set of concerns and criteria for studying them in a systematic way that fosters creative, satisfying, inclusive, and appropriate responses. A deal that has been cut to settle a dispute over government funding programs or academic turf is more likely to provide a laundry list of disparate items rather than a common vision and coherent agenda. A process of communal discernment guided by the Inward Light or, in more secular versions, by a commitment to the common good, might offer a more compelling and truthful way of defining the issues.

For a coherent agenda, it is important that the stakeholders from the relevant multiple constituencies and disciplinary paradigms enter into a dialogue in which they seek together a new framing of their concerns, one that speaks to each point of view but also speaks collectively as an integrated whole. Processes of communal discernment as practiced by Quakers and other traditions aim at something more than "mere consensus" that consists in acceptance of disparate items for disparate and perhaps even conflicting reasons, and requires an agreement to compromise.

Instead, Quakers aim at the discernment of a "unity" that is grounded in a common understanding and vision. For Friends, communal discernment seeks to bear witness to a Truth that might challenge deeply held beliefs and upset theories and social practices. This is something very different from mere consensus. To express the depth of this search, Quakers typically speak of concerns and responses to concerns, rather than problems and solutions.

The reasons for thinking that communal discernment would be useful in extra-paradigmatic research involving multiple disciplines apply at every scale of research. They also apply to much of the research done in natural sciences and most of the research dealing with social systems. They apply at every scale because power can be used to coerce agreement in the most intimate of settings as well as at the grandest of global scales and everywhere in between.

Communal discernment could be useful in much of the work done in natural sciences because so much of it requires collaboration among multiple disciplines in extra-paradigmatic contexts. Because social systems virtually always involve multiple dimensions that include economic, political, juridical, cultural, geographic, biological and other factors, extra-paradigmatic synthesis is required to describe, explain, and interact with them, which means that communal discernment would be a great asset.

The use of communal discernment does not guarantee that researchers will reach a unity of understanding and true conclusions. Communal discernment is understood here not as a substitute for scientific research, but as a way of making the character of its soundest traditions explicit and enhancing them in ways that promote open dialogue that aims at unity among participants and truth. There are no guarantees that even such enhanced processes will always work, but in the context of discovery, processes of communal discernment like those coming from the Quaker tradition are appropriate methods for researchers in any field of research.

In the context of justification, especially when extra-paradigmatic research is undertaken, processes seeking unity and truth offer a better prospect for arriving at coherent and reasonable ways of framing and dealing with research issues than the political processes that appeal to interests and power, or simply technical/bureaucratic approaches pursuing standard operating procedures rather than seeking unity and truth.

The shortcomings of the interest group and power politics approach are illustrated well by the approach to world hunger provided for decades by the U. S. Farm Bill, which provided subsidies for industrialized farming at the expense of the family farm.[40] Likewise, the shortcomings of a technical/bureaucratic approach are illustrated well by the Green Revolution response to hunger in India and elsewhere, which ended up putting small-holder farmers out of business and actually increased hunger.[41]

The suggestion that communal discernment could be applied broadly and with special force in the context of research on social reality is explored in more detail in the next section, which considers whether the teleological and other metaphysical assumptions that the Quaker process sometimes presupposes may be formulated so as to be compatible with post-Darwinian science.

Teleology in Quaker Communal Discernment

The Quaker process described in earlier sections includes frequent and central reference to experiences of "leadings," "voices," "callings," and "following the Light" in ways that seem to imply a teleological view of the process, something that is guided by an end or "telos" at which the process aims to arrive.

A skeptic might want to suggest the following: "In so far as Quaker or other forms of communal discernment are simply avoiding power plays and trying to promote the best dialogical aspects of traditional science, then that is well and good. But when they start to introduce talk of 'leadings' and 'voices' and feeling led to rise and 'speak out of the silence,' this in some way claims there is a higher spiritual power guiding their dialogue. This adds a set of very different and suspect elements to their research process."

Since Darwin, science has sought to describe and explain the world without reference to any such guidance from the Divine or any teleological understanding of reality as guided by purposes at all. This effort has come to include the description and explanation of science itself. Philosophers of science have sought to view science itself in post-Darwinian terms that allow what is thought of as "progress" to occur in something like the way the evolution of organisms occurs, as a process of unguided experiments or mutations that undergo natural selection and succeed or fail as a result.[42]

Teleological underpinnings of Quaker process, where they exist, may seem suspect or simply illegitimate for at least two reasons. First, there may be a kind of backward causation in time, as though the end at which the process arrives travels back through time to cause earlier events that then bring this end itself into being so that it causes itself before itself exists. This suggests there would be the time-travel and backwards-causation paradoxes familiar from science fiction (e. g. where someone turns out to be her own ancestor).

Second, if the causation is not backwards through time, then it would seem to involve intervention from some kind of mind or higher power outside of time, as in the teleological processes envisioned in miracles of divine creation of life forms and action in history. This would seem to run counter to the program of Darwinian theory which has sought to provide mechanistic ways of explaining apparent teleology and eliminate the use of all explanations that rely on any such divine interventions from outside time. Natural selection has explained the well-functioning and seemingly "purposeful" structures of organisms so well that there is a widespread commitment to rely on it and similar principles to explain away all apparent teleological phenomena.

Some theists may hold that miracles that violate the laws of nature can and do occur through divine intervention and this simply provides a limitation on the mechanisms of nature and the theoretical understanding of science. However, are there ways of understanding teleological processes in general—and those involved in communal discernment in particular—that do not require researchers to limit the realms of their study or assume laws of nature are sometimes invalid? In the short space allowed here it would not be possible to settle the issues at stake conclusively. But it may be possible to reframe them in a way that will allow respectful and fruitful discussions that give mainstream researchers reason to consider communal discernment seriously as an epistemological program worthy of pursuit. The vocabulary and philosophical ways of framing the issues proposed here are offered as one attempt to further such discussions.

Consider, for example, the analysis of mainstream biologists, such as, Ernst Mayr and Daniel Dennett, and how a shared frame for dialogue might be developed.[43] In introducing the relevant ideas, an initial analogy may help. Some ancient peoples thought that 1) unicorns roamed the earth, and 2) lightning bolts were hurled down from the sky by Zeus to send messages and destroy things. Science has proven both false, but in different ways. Research has shown in the first case that there simply are no unicorns. In the second case, research has shown that while lightning does in fact exist, it has a different cause, that of electromagnetic energy. In the case of apparent teleological processes in nature like the development of the flow of blood in animals, which is "for" all sorts of purposes including the circulation of oxygen, Darwinian science continues to affirm the development of blood flow, but offers an alternative explanation, natural selection over time, rather than divine intervention in an instant.

To articulate this distinction, writers such as Ernst Mayr have contrasted divinely caused *teleology* with the natural processes of "*teleonomy*" that can be explained with genetics and evolution. Mayr argues:

> *All teleonomic behavior is characterized by two components. It is guided by a "program" and it depends on the existence of some endpoint, goal, or terminus that is "foreseen" in the program that regulates the behavior process. This endpoint might be a structure (in development), a physiological function, the attainment of a geographic position (in migration), or a "consummatory act" ... in behavior. Each particular program is the result of natural selection, constantly adjusted by the selective value of the achieved endpoint.*
>
> *The key in the definition of teleonomic is the **genetic program**. The importance of recognition of the existence of programs lies in the fact that a program is both something material and something existing before the initiation of the teleonomic process. This shows that there is no conflict between teleonomy and causality. The existence of teleonomic processes regulated by evolved programs is the reason for the dual causations in biology, due to natural laws, as in the physical sciences, and due to genetic programs.*[44]

The use of such teleonomic explanations is a fundamental and pervasive feature of modern biology. It provides explanations for all sorts of functional, purposive behavior without appeals to deities. Teleonomic explanations refer to purposes and goals that guide the genetic programs, and the physical and behavioral adaptations that result. These programs are embodied in and enacted by physical molecules and organs, but they are codes, systems of information that require context and meaning to operate. They are embodied in a material substrate, but their defining features are the formal patterns that are so embodied. They are information, structure. This information is replicated and passed from one physical object to another, and thus the information exists beyond the original physical object. The "same" genetic program can be embodied in numerically distinct physical organisms, as in cell division (mitosis), which is the very stuff of life.

Design Space

To understand the process of evolution in which successive moments of natural selection result in preferential survival of some populations of organisms rather than others, it is helpful to appeal to some notion like the one Daniel Dennett refers to as "Design Space."[45]

The environments in which organisms evolve are defined not only by the three dimensions of physical space but also a host of other dimensions that define features of that space like temperature, humidity, salinity, the availability of food, wind speed, ease of travel, visibility, et cetera. The genetic programs of different organisms are competing for resources amidst the structures of that multidimensional space. As Dennett notes, that space is governed by the laws of logic and physics, and structured by the continually evolving and emerging facts of biology and history. These create the Design Space in which natural selection occurs. Its structures determine which organisms have better chances of successful reproduction and how the purposes and functions in their genetic programs will be selected out, successively, over time, to yield the historical developments natural selection explains.

The structures of Design Space let us understand all sorts of patterns in natural history. Honeycomb tubes are hexagonal in shape. If you want to pack tubes as strongly and densely as possible, a hexagonal shape is much better than square, pentagonal, septagonal or other shapes. This is a fact of geometry that is part of the Design Space in which bees evolved—and which created a bias in favor of the evolution of the bees that make such honeycombs.

Why have eye-like organs evolved multiple times from independent starting points in evolutionary history? The answer lies in physical facts about the Earth's environment in which the ability to detect and interpret electromagnetic radiation in the range of light provides enormously useful information that gives organisms survival and reproduction advantages. These features of Design Space constrain and guide the processes of natural selection. In doing so, they create historical facts that themselves further constrain and guide selection. So, for instance, once organisms have eyes, they enter a region of Design Space in which there are advantages to those who use color, shape and motion to signal potential mates—and competitors—for feeding, reproduction, and rearing activities. Clearly, if an organism evolved that had the capacity to detect structures in Design Space and guide its own behavior in light of them, it would have even more decided advantages. In the case of humans, this is what has occurred.

With the development of human languages, the purposeful and functional genetic programs studied by biologists are enriched with the purposeful and functional linguistic "programs" studied by anthropologists, economists and other social researchers. Much human behavior is not describable or explainable without the use of the interpretive categories of purpose, function, and intentionality. When we observe people in a market or a voting booth, we ask them what are they doing and why. We ask this of institutions they have established, like the market and the voting booth.

But much of human behavior is not guided by clear and fixed purposes and functions. It is exploratory and experimental in character. In the market we may encounter shoppers who are "looking for something that will plug a hole in the wall" though they are not sure what might serve that purpose best. Or they may be simply "browsing," in which case they may not even be sure of what their purposes or goals are, but they are open to discovering them as they explore the realms of Design Space represented in the things where they are shopping.

Adapting lives and improving behavior, individually and collectively, calls for the exploration not only of what are the facts on the ground in the present, but what are the possibilities in the Design Space of the future. Those possibilities are often difficult to discern in the midst of the noise of the present. The process can be very difficult and fraught with the irreducible uncertainties that emergence, non-linearity, chaos and other features of complex systems can involve. Along with the gift of linguistic abilities comes the power to center down and enter a silence that allows us to listen and look more deeply and attentively at the realms of possibility articulated in various fragments of our language "programs," and the hot spots or key points in those realms of Design Space where multiple values are realized in dramatically more rich, varied, fruitful, and resilient ways. The insights into those spots can provide "leadings" and the experiences of "voices, callings, and being led by the Light" that can guide exploratory research. Some of us may prefer to label such Design Space in what seems like more neutral metaphysical terms like "structures of possibility that are part of reality." Others may feel led, like Alfred North Whitehead, to describe it as the "Mind of God."[46]

Regardless of terminology used, Design Space provides an explanation of many natural and human processes. It is not a physical object but a set of forms or patterns that provide formal rather than material causes for explaining things. These formal causes define the patterns that programs in space and time use to define their goals or objectives and that guide natural processes which are teleonomic in character. Language makes it possible for humans to study Design Space through modeling portions of it in ways that allow for planning and the search for more intelligent, intentional behavior.[44]

Communal discernment as practiced by Friends may provide one useful way of discerning the structures of Design Space and truths about it. In so far as the formal content of the leadings and other teleological elements of the process are grounded in such structures, then the purposive character of the process does not involve backward causation in time of a suspect sort and makes use of notions already implicit in a post-Darwinian understanding of the world.

Having considered reasons for including teleological elements in understanding the world in general, questions remain regarding the claim that in research involving interpretations of human behavior, communal discernment processes are especially appropriate and likely to result in unified understanding and truth.

Communal Discernment in Social Research

How might research on human communities be different from the study of other organisms and their communities? Many philosophers and social researchers have considered this question at length.

One view of the issue largely dismisses it, holding there should be no fundamental differences between social research and natural science. In this view, humans are a part of nature and that the study of them should, in principle, be no different than the study of other natural phenomena. So, for instance, Carl Hempel and other philosophers in the Logical Positivist tradition have held that social scientists and historians should be using observation and experiment to discover laws of human behavior that can be expressed in mathematical formulas comparable to those of Newton's physics. They have held that social science should, like Newtonian physics, be descriptive rather than prescriptive. It should aim to simply describe what *is* rather than make value judgments about what *ought* to be.[47] Practicing psychologists like B. F. Skinner and social scientists like Milton Friedman have sought to pursue their research in these ways.[48]

Many practicing social scientists and historians have found it difficult, however, to pursue research on humans within the framework offered by this interpretation of the natural science tradition.[49] And many philosophers from both the Anglo-Saxon "Analytic" tradition as well as from the European "Continental" tradition have argued that there are systematic reasons why studies of humans must be pursued in ways that are importantly different from those of classical natural science. The terminology and the philosophical methods employed by these different critics of the Positivist tradition vary in some dramatic ways.[50] But they share commonalities including observations about the central roles and distinctive functions of language, intentionality, institutions, history and interpretation in understanding the lived world of humans and their communities.

In the Analytic tradition, Peter Winch and Ludwig Wittgenstein used analysis of language to investigate the distinctive features of social research.[51] They argued that the everyday languages humans speak structure social reality in ways that preclude being modeled by mathematical formulae in universal, a-historical laws. Instead social research requires an understanding that appeals to intentionality and the establishment of institutions. Interpreting human activity is more like reading a text than manipulating a mathematical formula. These basic points have been developed in a variety of ways by subsequent philosophers in the Analytic tradition as well as practicing anthropologists, historians and other social researchers.[52]

Philosophers working out of the continental European traditions have developed very similar core points, despite some dramatic differences in philosophical points of reference and methods. In that tradition, followers as well as critics of Martin Heidegger, Jacques Derrida, Michel Foucault, Jurgen Habermas, Simone de

Beauvoir and Donna Haraway include a diversity of thinkers who resist any attempt to categorize their thought collectively.[53] They articulate different ways for understanding the distinctive characteristics of human language, intentionality, institutions, history, and interpretation. But despite those differences, these thinkers do share an emphasis on the centrality of precisely those categories for understanding humans and how social research differs from the kind of work we associate with Galileo and his tradition of natural science.

One way to begin an exploration of the core insights coming out of these traditions and their relevance to understanding social research is by reflecting on the following: On the face of it, there seems to be a major difference between doing research on electrons or eels versus studying people. Studying why an electron jumps to a new energy level or a hawk migrates at a certain time requires evidence for hypotheses and this evidence is generated using observation and experiment. Studying why a person has purchased a Prius car or stepped into a polling booth can in a similar way make use of hypotheses, observation, and experiment. But human subjects can also be asked directly, "What are you doing and why?"

Unlike the electron or hawk, humans have a self-understanding of what they are doing, which is defined by purposes and values they are trying to realize, as well as background assumptions they make about the world and their immediate context. They may say something like, "I am buying a present for my wife in order to surprise her for her birthday," or "I am switching out to get a more energy-efficient vehicle and make use of the current tax rebates." Self-understanding is not something casually connected and detachable. A leg band on a hawk might reveal where it has come from. If the hawk could read it, it might tell her something as well. The band can be snipped off, the bird released, and she can continue on her migration without it. But shoppers or voters cannot perform their actions without a self-understanding of what they are doing. Their understanding is part of what constitutes the structure and nature of their action and connects it to the institutions of private property and democracy that provide an indispensable context for such actions.

This has very important implications for research on humans. First, it means that studying people requires learning their language and learning how they themselves would describe what they are doing and why. In studying physics, Galileo proclaimed that "Nature speaks mathematics" and argued that the language of formal algebra and geometry was necessary to describe and explain the motions of balls rolling down inclined planes. Analogously, people speak not only a variety of languages like Chinese and English, but they speak and structure their actions in terms of very specific and local varieties of languages that are associated with things like their local economic and political institutions, cultural practices and traditions. These are the things ethnographers study through participant observation and interview methods when they try to understand how, for example, a town meeting in eastern Maine is different from an *asamblea del ejido* in Yucatan. Such participatory research is a foundation for the study of human communities. Without it, no language exists to describe the basic phenomena researchers are trying to understand and explain.

The language and practices that structure such activity differ from the mathematical and mechanical descriptions and explanations that Galileo sought. Human languages involve words, phrases, and other "texts" that must be read or interpreted in a holistic way in the "con-text" of other texts. Languages are value laden and the values inform and provide "con-text" in pervasive ways. They do not lend themselves to value-neutral descriptions or explanations. The meanings of words and practices involve significant levels of convention that are instituted or established over time. Over time they are changed either through collective agreement, individual initiative, or community drift. Meanings of words and cultural practices are historical phenomena, quite unlike universal natural laws of physics. The structures of the concepts and practices that get instituted in these ways are typically pervaded by relationships that Wittgenstein called "family resemblance." There is not a single, neatly definable essence that runs through all the cases of the concept of a "game," "shopping," or "voting." Such language is not reducible to axiomatic expression the way a mathematical language is.

Participatory research may enable an understanding of the language and practices of a group, and the self-understandings that structure activities, but it is usually not an adequate endpoint for the study of human interaction for one basic reason: almost always self-understandings are incomplete in one or more ways. They may be vague, tacit, internally inconsistent, merely a partial understanding of what is going on, inaccurate in one or more of their beliefs about things, or they may fail to fully and effectively motivate the action for which they call.

What is called for in such cases is something that might be broadly considered a kind of *"critical participatory research,"* research that helps us create more adequate forms of self-understanding.[54] Those involved in a critical participatory research project may be assumed to share a common interest—both those who are members of the original community studying itself, as well as any other researchers who may join them as collaborators. They all want to know what is really being done and why. Everyone wants to revise the self-understandings involved in order to eliminate blind spots as much as is reasonably possible. For example, when shopping or voting, one would like to be clear, explicit, consistent, complete, and accurate to effectively motivate the intended action.

Any incoherency in understanding can create conflicts. Large incoherencies can create serious conflicts. Normally, one of the aims of social research is to improve understanding, mitigate incoherencies, and reduce conflict. In this way, critical participatory research that helps self-understanding is working to advance peace. Peacemaking is, in this respect, a central aim of social research. Communal discernment, as a form of collaborative, critical, participatory research, can be effective toward this end.

Do biased interests tend to distort research on human activities and social processes? Attempts to develop more objective and valid justifications for the claims of knowledge require greater consciousness of the way human interests bias investigations. In order to avoid partisan bias and advance holistic ways of knowing that include all the possible perspectives relevant to the phenomena under study, a high level of inclusiveness and democratic participation are needed. If this inclusive, egalitarian approach to objective knowing, and this justification of knowledge, seems promising, then Quaker practices might provide a useful model of how claims to knowledge could be developed and justified.

In summary, *participatory* research is important because social reality is constituted by the language and intentions of the participants in any given context of study. This language can only be learned by participation. *Critical* participatory research is necessary because the self-understanding of participants is often incomplete or flawed in other ways.

Critical participatory research that seeks to enhance self-understanding could be modeled on the five steps of Quaker process as noted in Chapter I. The vocabulary and customs used in critical participatory research may vary considerably, but the functions of quieting impulses, addressing concerns, seeking unity, finding clearness and bearing witness can facilitate research in the context of human social behavior.

At this point, an atheist or agnostic philosopher of science might argue: "It may make good sense to allow for these kinds of teleological considerations in the research process because it provides a sound way of making sense of the purposive character of scientific behavior. It may make sense to use critical participatory research methods that are analogous to aspects of the Quaker process of communal discernment. But what about the talk of the 'Inward Light'? Quakers seem to talk as though there is a kind of Divine Power that gives them special access to insights about Design Space and guides their version of critical participatory research towards truth. Beyond 17th century superstition, what could this involve?" The next section addresses this question.

The Inward Light, Intellect, and Imagination

Can the metaphysics or ontology that underlies Quaker process be framed in different traditions and languages? Do we find in other traditions an understanding of the process of human activity in research and the creation of knowledge that utilizes assumptions analogous to the Quaker view of reality? This section explores the ideas that 1) there is a power of illumination accessible to all persons that includes a capacity to reason in norm-informed ways; 2) these norms are grounded in trans-historical, objective, emergent truths; and 3) it is in the nature of our relationship to reality that leadings can be discerned to guide inquiry and action in ways that are informed by these trans-historical, objective, emergent truths.

Is the "Inward Light" merely a myth that helps Quakers act more collaboratively, or does it name some kind of power or Presence that exists objectively, that provides support and guidance in normative inquiry, and is accessible to all humans as humans—and not just to inculcated members of a peculiar sect? How can the nature of this reality be best understood?

Early Quakers interpreted their experiences of the Inward Light in the context of the Bible and a restoration of primitive Christianity. Their experience of that Light was understood as an experience that "Christ was come to teach people Himself, by his Power and Spirit in their hearts."[55] However, they also were quite insistent in their conviction that this Light was something that could be experienced directly, independently of any beliefs in or knowledge about Jesus of Nazareth or any of the writings in the Judeo-Christian tradition. They were firmly convinced that access to that experience and to the transforming power of the Presence of that Light was an essential endowment of all humans. As George Fox put it, there is "that of God" in everyone.

Contemporary Quakers likewise believe that this kind of transforming power or Presence is one that has been experienced in many different cultural and religious traditions, and is potentially available to everyone. It can, for example, inform everyone's participation in conflict resolution as they experience the love that makes peace between enemies. This is a transforming power that can redefine ourselves, the situations we are in, and the options available for us and those previously thought of as enemies.

How might someone coming out of a non-Quaker tradition and language community try to make sense of talk about the Inward Light? One way to frame some common language for dialogue about this might be to draw on the concepts of "Design Space" and dialogue.

The previous discussion of teleological phenomena noted that the world includes not just physical objects in space and time but also patterns in Design Space including systems of meaning and characteristic forms and relationships. These realities provide part of the explanation of many processes in the world. They help explain the structural features and the dynamic behavior of things, their formal causes.

A question arises about creatures with the kind of language abilities humans have: How in their thinking, imagining, creating, planning, and deciding, do we relate to the forms that are possible but not yet actual? How, in our dialogues, should we relate to the potentialities in Design Space that we hope our actions will bring about (or prevent) in Real Space? The actual language or physical modeling tools that are used in the present moment in space and time might embody features of possible Designs. For example, an architect builds a scale model of the house she plans, or a non-violent activist role-plays a possible encounter with the police. The same basic form might be present in the model and the building, or in the role-play and actual police encounter. What is the power that enables us to see such forms and project them in our thinking about the future?

Whether it is called "imagination," "intuition," "intellect," or "reason," it is clear that humans have the ability to know such forms and make considered judgments about them. In a related way, people generally have the ability to put themselves in another's shoes. The concept of the Inward Light is not necessary to navigate the potentialities of Design Space or to experience how others feel. Whatever image is adopted, whatever language is called forth to identify this human endowment, its activation, cultivation, and practice is central to our relationship with the reality in which people function. Activating, cultivating, and practicing this potentiality, this access to Inward Light, is what gives communal discernment its unique and helpful place in social research.

Most of the forms of concern are intimately connected to core values, purposes, interests, principles, functions, *etc*. For example, activities and products of architecture and non-violent resistance only make sense in the context of intentions and institutions informed by values that define what houses are for, or what social change movements seek. Research on human communities involves studying how they discern and act in light of their understanding of where they are and what they are doing. This discerning is always in relation to the Design Space of possibilities that provide the formal patterns they might realize in the future. These are all value laden and value "in-formed" potentialities that they are discerning.

Now, suppose researchers from one cultural setting are studying people who inhabit a different cultural tradition. The researchers may have a variety of ideas and values that do not map onto the mental world, the values, and behavior of the people they are studying. In the extreme case, the "Other" that the researchers are studying might be perceived as an enemy. How do researchers go about understanding cultures significantly different from their own? First, they need to bracket a good deal of the ideas and values that give form to their own actions and practices. They need to somehow quiet or silence the assumptions they would normally make about things so as to allow themselves to step beyond their region of Design Space and into the region of Design Space inhabited by the Other.

Besides bracketing the forms that structure familiar regions, researchers also need to embrace or affirm, at least initially, the forms of the Other. If they don't, they won't be able to see the logic of things as the Other thinks about them or see the world through their eyes. The researchers need, in other words, to adopt a stance that can be described as empathic. Is it possible to be fully empathic without going a step further and being affirming or loving or compassionate? Perhaps. The shift to seeing the world through the eyes of the Other is certainly facilitated strongly by outright compassion and affirming of the Other.

To engage with the Other as a member of a community, sharing a common collaborative process, such affirmation may in general be a prerequisite the Other would impose before being ready to acknowledge and accept a researcher as a collaborator. Knowing who this Other is requires engaging in the kind of participatory research that begins with empathic listening and observing and moves through nuance and correction to collaborative, critical participation involving compassion.[56]

Participation in this kind of research will typically lead to changes in how researchers see and think about themselves. The researcher's own beliefs, values or practices may be called into question and she may in fact be led to revise them. This process of participatory observation can be transformational for the observer. It can enlarge one's sense of self to include one's core identity as associated with a friend, family, group, or community of a larger sort.

What is this power to bracket the self and its cultural context, affirm the Other, and engage in the kind of relationship-building and peacemaking associated with collaborative, critical participatory research? It is not itself a mere physical object or thing in the world any more than the forms or peoples it studies are. The history of Western Philosophy is, in part, the history of a series of attempts to find language to describe and account for this power and the processes of understanding and dialogue it makes possible.

In the Classical Period thinkers like Plato would articulate this realm as "intellect" or "nous" and a power of love as "Eros" that led people to seek and know Forms like the Ideas of Beauty and Justice.

In the Modern Period, philosophers like Descartes and Kant developed accounts of imagination, reason and the power of synthesis in its theoretical and practical employments to articulate this realm and these activities.

Philosophers in the more contemporary Analytic Tradition, like W. V. O. Quine, might articulate this realm and these kind of activities by saying they occur not in the "object language" we use to describe the world in our theories of it but in the "meta-language" or "background language" we use to connect it to the world and to revise it. In this language we use indexicals like "here" and "now" that assign reference relations.

In the Continental Tradition growing out of Husserl and Heidegger, different vocabularies characterize the processes through which humans refer to the world, live in the present, and enter into dialogue with each other. Successive philosophers in that tradition have argued that it is in the very nature of these processes that any attempt to understand them must always remain problematic.

Martin Buber's analysis in *I and Thou* may provide the most succinct and accessible way of framing the difficulties that arise in trying to articulate the key distinctions at issue here. He notes that people who enter dialogue with each other, listening, critiquing, judging, changing their beliefs and practices are not mere things which have simply an "I-It" relationship.[57] They are persons who can call themselves into question and critique their entire view of the world with respect to a deeper engagement with the reality of the Other. They are inevitably called to enter into "I-Thou" relations, treating the Other as a You with whom they might enter into agreements as a We. And the power of relating involved here takes place in the realm of meanings, as much as in the realm of the objects that give them partial embodiment. The realm of meanings is lived in time, here and now, in the flow of the ever-moving Present.

As Buber and others have noted, a variety of philosophically perplexing questions come up in trying to give an account of such I-Thou relations and the experience of Presence in the lived moment. For example, attempts to define the I-Thou relation itself seem highly problematic because they unavoidably treat the relationship as an "It" and create an "I-It" relationship. The I-Thou experience would seem to have slipped away or become obscured. It would seem that while it might be something that can be experienced, it cannot, in that sense, be defined—by definition, so to speak.

Further, philosophical attempts to analyze the experience of the lived moment as a Presence or an ongoing Present have proved frustrating and perplexing in a variety of ways for philosophers since the time of Zeno. Is the Present moment an instant with no temporal length whatever, in which case when is anything really present? Or does it consist of some minimum quantum of time in each successive Present? If so, when does it move from one present to the next?

Like Zeno and his mentor, Parmenides, many interpreters of modern Einsteinian physics hold that the experience of a moving Present is actually an illusion. The Special Theory of Relativity would seem to tell us that time is simply a fourth dimension of a space-time manifold in which there are different locations, but no single moving present of universally simultaneous experience moving from the past to the future.

In that view, the ordinary experience of the Present is as much an illusion as is the pre-Copernican perception that in the morning the Sun "rises" in the sky.[58] And yet . . . you do seem to be here, now, reading this at this moment, don't you? Augustine's classic statement about the puzzles concerning time echoes the point just made about I-Thou relations, that they are vital and real experiences and yet elude logical definition. As Augustine put it, "What then is time? If no one asks me, I know; if I wish to explain it to one that asketh, I know not."[59]

What these sorts of philosophical difficulties suggest is perhaps a need to acknowledge that there are in fact fundamental aspects of experience that are quite real and of central importance in understanding human life, but which elude the kind of definitions sought in mathematics, logic, and natural science. Quakers have traditionally approached such experiences with an empiricist temperament, acknowledging the reality of what is experienced as well as the difficulty in capturing it in words. This is the basis of their non-creedal approach to their core testimonies and their grounding of their talk about "Inward Light," not in metaphysical abstractions, but in shared experiences within the contexts of meetings for worship, silent meditation, and dialogue.

This experiential and non-creedal approach can be contrasted to Enlightenment concepts of rationality which have been widely critiqued by Post-Modernists.[60] In the work of Immanuel Kant, Jeremy Bentham and other Enlightenment thinkers, there is an assumption that Reason is a faculty or power that is uniform and universal in nature, and that scientific and ethical truths can be determined according to its principles using a monological process. That is, an enlightened independent thinker can determine the laws of motion that govern the universe or the laws of morality that should govern society. The Quaker interpretation of experiences of "Inward Light" is quite different from this view of "en-Light-enment" because it supposes that there is no single, absolute, essential, universal form of reason. Instead, principles and practices of reasoning can take a variety of forms.

This latter view does not imply, however, that the different forms must be simply accepted as competing, incompatible principles or practices. Instead of adopting a relativist view of such differences, Quaker practice is to seek openings for agreement between them in processes of dialogue. In the realm of ethics, this takes the form of what might be called a "Rainbow Version" of the "Golden Rule." Instead of proposing that people should "Do unto others as you would have them do unto you," Quaker practices of deep listening invite us to "Do unto others as they would have us do unto them."

The Quaker experience is that rational beliefs about the world, and wise choices for acting in it, are not arrived at by simply following one's own assumptions and observations through engaging in a monologue that is self-consistent. Instead, dialogue with others is necessary. In this respect, the role of silence in Quaker process is not to shut out the world but to help quiet our inner monologue so that we can listen more deeply and enter more profoundly into dialogue with others. The experiences Quakers are trying to describe in talk about the Inward Light are not experiences of a Light of Reason that reveals fixed essences populating an a-historical system of definitons and deductions. They are experiences of open, dynamic, growing insights into realities characterized by emergence.

Entering into dialogue with others in the lived moment and being Present with them as people rather than things are central to the experience of being human, participating in communities and engaging in research. The activity of research within human communities presupposes that there is some power that enables one to enter in imaginative and discerning ways into the "worlds" of Others. To affirm the existence of such a power and to speak candidly of experiences of it, especially the more vivid and intense experiences, may lead some people to feel they have entered a realm that is in some sense unscientific, subject to superstitious belief or "metaphysical" and "spooky" in perjorative senses of those terms. They may wish to resist terms for describing such experiences that make strong associations with religious traditions. As long as this encounter leads to more clarity about meaning and descriptions of what people actually experience, perhaps the power in question will not seem any more "spooky" than the forms it enables or the I-Thou dialogue between persons it makes possible.

While Quakerism has made the experience of the Inward Light a central feature of its practice, many other people also find the concept and its religious associations helpful. The fact that corresponding experience and similar metaphors of description have arisen in different traditions provides an historical context of language and practice for activities of self-transcendence, love, and compassion that facilitate the process of creating a more communal world. Quaker experience and the language of the Inward Light provides a context that opens the process of communal discernment into broad avenues of social research application.

Summary

The aim of this chapter was to develop the initial elements of a language for describing and analyzing the philosophical presuppositions of Quaker process in a way that makes it available for consideration and possible adoption by Quaker researchers, practitioners of other forms of communal discernment, and by mainstream secular researchers as well. Further, it is hoped that researchers may see a potential benefit in utilizing these presuppositions and practices, translating them into language that is appropriate for the tradition out of which their own work grows. Further dialogue about these issues can enrich and improve both Quaker and non-Quaker approaches to research.

Communal discernment methods are useful in the context of discovery. In the context of justification, especially when research extends outside of some single, stable paradigm, communal discernment offers a method for seeking coherent and reasonable ways of framing and dealing with research issues. This approach makes the case that seeking unity and truth may be a better decision-making platform for social research than political processes that appeal to interests and powers.

In describing and explaining many natural and human processes it may be helpful to appeal to formal causes that are patterns defining the Design Space, which is the structure of possibilities within which the "programs" of biological organisms develop their goals through teleonomic processes. These structures can be studied through modeling, planning, and other forms of intelligent, intentional behavior.

Studies within human communities take the form of critical participatory research because social reality is constituted by the language and intentions of the participants involved. This language can only be learned by participation. However, the self-understanding of participants is typically flawed in multiple ways that may include being tacit, vague, inconsistent, inaccurate, incoherent and/or failing to adequately express or motivate the action it claims.

Critical participatory research utilizing communal discernment can help create a more complete and accurate self-understanding for all participants. The five steps typical of Quaker communal process provide a well-tested model. In practice, language may vary considerably depending on the setting, but the functions involved in letting go of impulses, addressing concerns, seeking unity, finding clearness, and bearing witness could be integral parts of any critical participatory research method.

In dialogue, and in research on human communities in particular, we generally presuppose access to a power that enables us to enter into "I-Thou" relationships in a realm of meanings experienced in a lived present. In these relationships we may experience both group and self-transformation as part of the process of shared inquiry. This power has the functional features that characterize what Quakers refer to as the "Inward Light." While reasonable people may disagree about how to explain the ultimate sources or undergirding of this power, it is not hard to recognize that we share some such power, and that it is because of this sharing that we are able to enter into critical participatory social research that results in authentic and trustworthy knowledge.

CHAPTER V
Closing

The record of human impacts on the world and the problems created is humbling. Research in biodiversity shows that the sixth great extinction of life in planetary history is underway and being caused almost entirely by human impacts. An estimated 10,000 species are being lost annually. At the same time, over two billion people continue to live on less than two dollars a day and economic inequality is intensifying. These realities are integrally connected to the continued development of massive military expenditures and to the interventions of massive economic development that is altering the face of the earth and the structures of its human and other natural communities. In response, we are called as individuals and communities to do something—to do a lot. And to do it now.

Yet we are also called to take pause, to reflect, and to be humbled in a radical way by our lack of understanding. We are called to assess our collective roles in these complex, interconnected problems that often involve exponential rates of change. It is clear that often the very problems of concern are the results of well intentioned efforts by the best and the brightest who were relying on the most advanced theories of chemistry, agriculture, economics, sociology, and other disciplines available.

Where then do we turn? A key step towards the humility we need to live rightly in the world involves ceasing the affirmation of our current beliefs and opening them up to question. We must allow ourselves the time to stop talking and start listening. This involves deep listening that demands that we begin by stopping the constant stream of impulses, desires and beliefs that babble through us and that we enter an open and deepening silence. It is the experience of Quakers that entering such silence collectively can be mutually empowering; it can help everyone present enter more deeply and listen with greater care and discernment. It is the Quaker experience that the stillness of such silence does not abandon us to doubt or leave us at a loss. In seeking amidst the dark shadows of our problems and concerns, we find there is some light that casts those shadows. The very difficulties over which we agonize arise for us precisely because we have some at least tacit, discernible sense of the gap between the way things are and the way they would be if they were made better. The practice of that discernment can be painful and difficult, but it can also lead to a world in which the shadows cast shorten and the light of day illuminates things from all sides in an increasing clarity.

In considering how to do research that may offer us the understanding we need to respond to urgent calls to better our world, we have much to learn about how to practice humility, enter silence, and use collective discernment. Quaker and other traditions offer resources for our continued experiments with communal discernment and collaborative research, and assurance that in the silence we can have openings, not only of seeking, but of finding as well. Through continuing revelation, a Presence works through us and comes to offer leadings, guidance, wisdom and a powerful transforming love.

Endnotes

1. Readers interested in seeing the results of these efforts can turn to the archive of work provided on the QIF website <quakerinstitute.org>.

2. Friends of different traditions describe themselves and their inquiries in different ways. The Quaker Institute for the Future sponsors Circles of Discernment that draw on the traditional Quaker practice of unprogrammed Meetings based on silence. Monthly Meetings that carry on this practice of discernment and guidance are generally associated with Friends General Conference. Other branches of the Quaker tradition are rooted in a more explicit commitment to a Christ-centered consideration of core Quaker values and gather for programmed worship services of a more conventional sort. Friends United Meeting (FUM) "commits itself to energize and equip Friends through the power of the Holy Spirit to gather people into fellowships where Jesus Christ is known, loved and obeyed as Teacher and Lord." <fum.org/about-us>. FUM believes that the continuing revelation to which Friends subscribe should be "consistent with the Scriptures, because the Holy Spirit is the source of both" <fwccworld. org/kinds.html>. We recognize that the degree to which Friends of different traditions commit to a Christ-centered orientation may impact how they might practice and bear witness to Meeting for Worship for the Conduct of Research. We hope Friends of different traditions can come together in a practice of discernment that yields insight into a new ethical understanding of the human-Earth relationship grounded in and informed by contemporary science.

3. Britain Yearly Meeting's Quaker Faith and Practice <quakerweb.org.uk/qfp>; Philadelphia Yearly Meeting's Faith and Practice <pym.org/faith-and-practice>; Quaker Information Center "Online Books of Faith and Practice" <quakerinfo.org/quakerism/faithandpractice>. Studies of Quaker practice include: Sheeran,1983; Brinton, 1993; Loring, 1997; Cox, 1985; Joy, 2011, pp. 50-56; Lacey, 2003.

4. Hill, 1984.

5. Declaration to King Charles II of England in 1660 by Quaker Margaret Fell. Note that in 1660 when this was written, "virtue" was understood to mean "strength."

6. Some of the material in this section is adapted from Cox,1985.

7. Sheeran, 1983, p. 81.

8. Fox, 1997.

9. Searle, Stanford J., Jr. "Embodied Knowing and Quaker Silence," submitted to Pendle Hill Pamphlers.

10. Belenky, *et al.*, 1997 and Jaggar, 1997.

11. Cox, 1985.

12 Ferguson and Rinard, 1999.

13 Parker Palmer <couragerenewal.org/parker/writings/clearness-committee>.

14 The Listening Project <listeningproject.info/how.php>.

15 Boulding, 1990, p. 110 ff.

16 Alternatives to Violence Prject <avpusa.org>.

17 Ambler, 2002.

18 Quakers were called "Friends of the Truth" when the Religious Society of Friends was founded in the middle of the 17th Century.

19 Fox, 1997.

20 Repko, 2011; Moulton, 1983; and Lakoff, 2004.

21 The material included here is from a forthcoming essay by Stanford J. Searl, Jr., on "Clearness Committees and Quaker Community." The clearness process it describes was associated with his work on two books: *The Meanings of Silence in Quaker Worship.* Lewiston NY: The Edwin Melen Press (2006); and *Voices from the Silence.* Bloomington IN: Authorhouse (2005).

22 Boulding, 1990.

23 Holliday, 2006.

24 Fox, 1997, p. 263

25 Brown, 2009, p.212.

26 Woolman, 1971.

27 *Ibid.*

28 Excerpt from *The Man Who Killed the Deer* by Frank Waters, © 1943, Frank Waters. This material is used by permission of Ohio University Press <ohioswallow.com>.

29 *Ibid.*

30 *Ibid.*

31 Joy, 2011, pp. 59 ff.

32 Coover *et. al.*, 1985.

33 Louis, 1994.

34 Gould, 1990.

35 For example, the Delphi method was developed for anonymity in communication to enhance communication between experts collaborating in forecasting and other studies of the future <en.wikipedia.

org/wiki/Delphi_method>. The Millenium Project uses the Delphi method in "an independent non-profit global participatory futures research think tank of futurists, scholars, business planners, and policy makers who work for international organizations, governments, corporations, NGOs, and universities. The Millennium Project manages a coherent and cumulative process that collects and assesses judgments from over 3,500 people since the beginning of the project selected by its 49 Nodes around the world" <millennium-project.org>.

36 National Science Foundation <nsf.gov>.

37 Innes and Booher, 2010, p. X. Another important tradition of interdisciplinary study and collaborative research has been developed by participants in the Society for Human Ecology. For an overview of their history and progress in this, see Borden, 2008.

38 Repko, 2011.

39 Kuhn and Hacking, 2012.

40 Gilbert Gaul, Sarah Cohen, and Dan Morgan. Federal Subsidies Turn Farms into Big Business. *Washington Post,* December 21, 2006 <washingtonpost.com/wp-dyn/content/article/2006/12/20/AR2006122001591.html>.

41 Pingali, 2012.

42 The precise formulations of this have varied significantly but versions of this approach appear in all of the pragmatists (Peirce, Dewey, James) and many more contemporary philosophers in the "Analytic" tradition such as W. V. O. Quine and Daniel Dennett.

43 Mayr, 2007, and Dennett, 1996.

44 Mayr, 2007, pp. 52-3.

45 Dennett, 1996

46 Whitehead, 1969

47 Hempel, 1948; Hempel, 1966.

48 Kenneth Boulding's *The Image: Knowledge in Life and Society.* Ann Arbor MI: University of Michigan Press, 1956

49 Geertz, 1977; Spradley, 1979; Charmz, 2006; Collingwood, 1994; Mink, 1966.

50 Cusset, 2008; Schrift, 2005.

51 Winch, 2007.

52 Delanty and Strydom, 2003; Fay, 1996.

53 Gadamer, 2013; de Beauvoir, 2011; Merleau-Ponty, 2003; Ricoeur, 1981; Derrida, 1998); Lacan, 2007; Foucault, 1984; Habermas, 1972;

Freire, 2000; Haraway, 2003; Robert Chambers' Rural Development: Putting the Last First, Routledge, Kindle Edition, 2014.

54 "Critial participatory research" is meant in the general terms articulated here. For more details see Cox, 1986 (chapter four), and Richard, 1984. A variety of authors use related terminology to articulate somewhat related views including "action research," "community-based research," "participatory research," "grounded theory," "critical theory" and "elicitive" research (Ledearch', 1995).

55 Fox, 1651-52. See also Penn, 1696.

56 Powdermaker, 1966.

57 Buber, 1971.

58 Grünbaum,1963.

59 St. Augustine, *The Confessions*, Book VI, <sacred-texts.com/chr/augconf/aug11.htm>. For more on time, see Heidegger, 2010.

60 For a general overview of Post-Modernism as it relates to social research, see Rosenua, 1992. Readers familiar with her work should note that the forms of Post-Modernist thought with which Quaker practice has perhaps the most affinity are those which she refers to as "Affirmative" rather than "Skeptical" Post-Modernists. Good examples of such "Affirmative" versions can be found in Esteva and Prakash,1998; and Esteva, 1987.

Bibliography

Ambler, Rex, 2002. *Light to Live By.* London: Britain Yearly Meeting <quakerbooks.org/light_to_live_by.php>.

de Beauvoir, Simone, 2011. *The Second Sex.* New York NY: Vintage.

Belenky, Mary, et. al., 1997. *Women's Ways of Knowing.* New York NY: Basic Books.

Borden, Richard J., 2008. A Brief History of SHE: Reflections on the Founding and First Twenty Five Years of the Society for Human Ecology, in *Human Ecology Review* 15 (1):95-108 <humanecologyreview.org/pastissues/her151/borden.pdf>.

Boulding, Elise, 1990 *Building A Global Civic Culture.* Syracuse Studies on Peace and Conflict Resolution, Syracuse NY: Syracuse University Press.

Boulding, Kenneth, 1956. *The Image: Knowledge in Life and Society.* Ann Arbor MI: University of Michigan Press.

Brinton, Howard, 1993. *Friends for 350 Years.* Wallingford PA: Pendle Hill Publications.

Brown, Peter G., Geoffrey Garver, Keith Helmuth, Robert Howell, Steve Szeghi, 2009. *Right Relationship: Building a Whole Earth Economy.* San Francisco CA: Berrett-Koehler Publishers.

Buber, Martin, 1971. *I and Thou.* New York NY: Touchstone.

Chambers, Robert, 2014. *Rural Development: Putting the Last First.* New York NY: Routledge.

Charmz, Kathy, 2006. *Constructing Grounded Theory: A Practical Guide through Qualitative Analysis.* Thousand Oaks CA: SAGE Publications Ltd.

Collingwood, R. G. , 1994. *Philosophical Study of the History of Methods in History*, Revised edition. New York NY: Oxford University Press USA.

Coover, Virginia, Ellen Deacon, Charles Esser, Christopher Moore, 1985. *Resource Manual for a Living Revolution.* Philadelphia PA: New Society Publishers.

Cox, Gray, 1985. *Bearing Witness: Quaker Process And A Culture Of Peace.* Wallingford PA: Pendle Hill Pamphlets.

Cox, J. Gray, 1986. *The Ways of Peace: A Philosophy of Peace as Action.* Mahwah NJ: Paulist Press.

Cusset, Francois, 2008. *French Theory: How Foucault, Derrida, Deleuze, & Co. Transformed the Intellectual Life of the United States.* Minneapolis MN: Univ of Minnesota Press.

Delanty, Gerard and Piet Strydom, 2003 *Philosophies of Social Science: The Classic and Contemporary Readings*. New York NY: Open University Press.

Dennett, Daniel, 1996. *Darwin's Dangerous Idea*. New York NY: Simon and Schuster.

Derrida, Jacques, 1998. *Of Grammatology*, Corrected edition. Baltimore MD: Johns Hopkins University Press.

Esteva, Gustavo, 1987. Regenerating People's Spaces in *Towards a Just World Peace: Perspectives from Social Movements*, Saul Mendelowitz, ed. Oxford UK: Butterworth-Heineman.

Esteva, Gustavo and Mahdu Suri Prakash, 1998. *Grassroots Postmodernism: Remaking the Soil of Cultures*. London UK: Zed Books.

Fay, Brian, 1996. *Contemporary Philosophy of Social Science: A Multicultural Approach*. Hoboken NJ: Wiley-Blackwell.

Ferguson, Janet, and Janet Rinard, 1999. *As Way Opened: A History of Atlanta Friends 1943-1997*. Atlanta GA: Atlanta Friends Meeting.

Foucault, Michel, 1984. *The Foucault Reader*. NewYork NY: Pantheon.

Fox, George, 1997. *The Journal of George Fox*. John L. Nickalls, ed. Philadelphia PA: Philadelphia Yearly Meeting of Religious Society of Friends.

Fox, George, 1651-52, *Journal*, Chapter VI, <strecorsoc.org/gfox/ch06.html>.

Freire, Paulo 2000. *Cultural Action for Freedom*. Cambridge MA: Harvard Educational Publishing Group.

Gadamer, Hans-Georg, 2013. *Truth and Method*. London UK: Bloomsbury Academic.

Geertz, Clifford, 1977. *The Interpretation of Cultures*. New York NY: Basic Books.

Gould, Stephen Jay, 1990. *Wonderful Life: The Burgess Shale and the Nature of History*. New York NY: W. W. Norton & Company.

Grünbaum, Adolph, 1963. *The Philosophical Problems of Space and Time*. New York NY: Alfred A. Knopf.

Haraway, Donna. *The Haraway Reader*. New York NY: Routledge.

Heidegger, Martin, 2010. *Being and Time: A Revised Edition of the Stanbaugh Translation*. Albany NY: State University of New York Press.

Helmuth, Keith, Judy Lumb, Sandra Lewis, and Barbara Day, 2006. Changing World View and Friends Testimonies. *Quaker Eco-Bulletin*

6(4) <quakerearthcare.org/sites/quakerearthcare.org/files/qeb/qeb-6-4-eco-theology-final.pdf>

Hempel, Carl, 1966. *The Philosophy of Natural Science*. London UK: Pearson.

Hempel, Carl, 1948. The Function of General Laws in History, in Carl G Hempel & Paul Oppenheim, eds. "Studies in the Logic of Explanation," *Philosophy of Science* 15(2):135–175.

Hill, Christopher, 1984. *The World Turned Upside Down: Radical Ideas during the English Revolution*. New York NY: Penguin.

Hoyle, Fred, and N. Chandra Wickramasinghe, 1981. *Evolution from Space*. London J.M. Dent & Sons.

Holliday, Laura, 2006. The Global Transformation of Corporations, Financial Institutions, and Government: A Quaker Approach. <quakerinstitute.org/wp-content/uploads/2012/12/FGC_2006_lecture.pdf>.

Habermas, Juergen, 1972. *Knowledge and Human Interests*. Beacon Press.

Innes, Judith E. and David E. Booher, 2010. *Planning with Complexity: An Introduction to Collaboarative Rationality for Public Policy*. New York NY: Routledge.

Jaggar, Allison, 1997. Love and Knowledge: Emotion in Feminist Epistemology. In *Feminist Social Thought: A Reader*, Diana Tiejens Meyer, ed. Routledge, New York.

Joy, Leonard, 2011. *How Does Societal Transformation Happen? Values Development, collective Wisdom, and Decision Making for the Common Good*. Quaker Institute for the Future Pamphlet 4. Caye Caulker, Belize: Producciones de la Hamaca.

Kuhn, Thomas, and Ian Hacking, 2012. *The Structure Of Scientific Revolutions: 50th Anniversary Edition*. Chicago: University of Chicago Press.

Lacan, Jacques, 2007. *Ecrits: The First Complete Edition in English*. New York NY: W. W. Norton & Company.

Lacey, Paul, 2003. *The Authority of Our Meetings is the Power of God*, Wallingford PA: Pendle Hill Pamphlets.

Lakoff, George, and Mark Johnson, 2003. *Metaphors We Live By*. Chicago IL: University of Chicago Press.

Ledearch, John Paul, 1995. *Preparing for Peace: Conflict Transformation across Cultures*. Syracurse NY: Syracuse University Press.

Loring, Patricia, 1997. *Listening Spirituality: Volume I, Personal Spiritual Practices Among Friends*. Washington DC: Open Press.

Louis, Meryl Reis, 1994. *Journal of Organizational Change Management* 7 (1): 42-60.

Mayr, Ernst, 2007. *What Makes Biology Unique? Considerations on the Autonomy of a Scientific Discipline.* Cambridge UK: Cambridge University Press.

Merleau-Ponty, Maurice, 2003. *Basic Writings.* New York: Routledge.

Mink, Louis O,1966. The Autonomy of Historical Understanding. *History and Theory*, 5 (1): 24–47.

Moulton, Janice, 1983. A Paradigm of Philosophy: The Adversary Method, pp. 149-164, in *Discovering Reality: Feminist Perspectives on Epistemology, Metaphysics, Methodlogy and Philosophy of Science*, Sandra Harding and Merrill B Hintikka, eds. Boston MA: Reidel.

Penn, William, 1696. *Primitive Christianity Revived.* <www.strecorsoc.org/penn/title.html>.

Pingali, Prabhu, 2012. Green Revolution: Impacts, Limits, and the Path Ahead. PNAS 109 (31): 12302-12308 <www.pnas.org/content/109/31/12302.full>.

Powdermaker, Hortense, 1966. *Stranger and Friend: The Way of the Anthropologist.* New York: W.W. Norton.

Repko, Allen F., 2011. *Interdisciplinary Research: Process and Theory*, Second Edition. Thousand Oaks CA: SAGE Publications, Inc.

Richard, Howard, 1984. *The Evaluation Of Cultural Action.* Basingstoke UK: Palgrave Macmillan.

Ricoeur, Paul, 1981. *Hermeneutics and the Human Sciences: Essays on Language, Action and Interpretation.* Cambridge UK: Cambridge University Press.

Rosenua, Pauline Marie, 1992. *Post-Modernism and the Social Science: Insights, Inroads and Intrusions.* Princeton NJ: Princeton University Press,

Schrift, Alan D., 2005. *Twentieth-Century French Philosophy: Key Themes and Thinkers*, 1st edition. Wiley-Blackwell.

Sheeran, Michael, 1983. *Beyond Majority Rule: Voteless Decisions in the Religious Society of Friends.* Philadelphia PA: Philadelphia Yearly Meeting of Religious Society of Friends.

Spradley, James P., 1979. *The Ethnographic Interview.* New York NY: Harcourt, Brace, Jovanovich.

Waters, Frank, 1943. *The Man Who Killed the Deer.* Denver: Sage Books, Alan Swallow.

Waters, Frank, 1963. *Book of the Hopi.* New York, Viking Press.

Whitehead, Alfred North, 1969. Process and Reality, New York NY: The Free Press (MacMillan Publishing),

Winch, Peter, 2007. *The Idea of a Social Science and Its Relation to Philosophy*, 1st edition. New York NY: Routledge.

Woolman, John, 1989. *The Journal and Major Essays of John Woolman*, Phillips P. Moulton, ed. Richmond IN: Friends United Press.

QUAKER INSTITUTE FOR THE FUTURE
Advancing a global future of inclusion, social justice, and ecological integrity through participatory research and discernment.

The Quaker Institute for the Future (QIF) seeks to generate systematic insight, knowledge, and wisdom that can inform public policy and enable us to treat all humans, all communities of life, and the whole Earth as manifestations of the Divine. QIF creates the opportunity for Quaker scholars and practitioners to apply the social and ecological intelligence of their disciplines within the context of Friends' testimonies and the Quaker traditions of truth seeking and public service.

The focus of the Institute's concerns include:
- Economic behavior that increasingly undermines the ecological processes on which life depends.
- The development of technologies and capabilities that hold us responsible for the future of humanity and the Earth.
- Structural violence and lethal conflict arising from the pressures of change, increasing inequity, concentrations of power and wealth, declining natural capital, and increasing militarism.
- The increasing separation of people into areas of poverty and wealth, and into social domains of aggrandizement and deprivation.
- The philosophy of individualism and its socially corrosive promotion as the principal means for the achievement of the common good.
- The complexity of global interdependence and its demands on governance systems and citizen's responsibilities.
- The convergence of ecological and economic breakdown into societal disintegration.

QIF Board of Trustees: Gray Cox, Elaine Emmi, Phil Emmi, Geoff Garver, Keith Helmuth, Laura Holliday, Leonard Joy, Judy Lumb, Shelley Tanenbaum, and Sara Wolcott. <*quakerinstitute.org*>

CPSIA information can be obtained
at www.ICGtesting.com
Printed in the USA
FFOW05n1058280514

9 789768 142573